Performance Testing of Textiles

The Textile Institute and Woodhead Publishing

The Textile Institute is a unique organisation in textiles, clothing and footwear. Incorporated in England by a Royal Charter granted in 1925, the Institute has individual and corporate members in over 90 countries. The aim of the Institute is to facilitate learning, recognise achievement, reward excellence and disseminate information within the global textiles, clothing and footwear industries.

Historically, The Textile Institute has published books of interest to its members and the textile industry. To maintain this policy, the Institute has entered into partnership with Woodhead Publishing Limited to ensure that Institute members and the textile industry continue to have access to high calibre titles on textile science and technology.

Most Woodhead titles on textiles are now published in collaboration with The Textile Institute. Through this arrangement, the Institute provides an Editorial Board which advises Woodhead on appropriate titles for future publication and suggests possible editors and authors for these books. Each book published under this arrangement carries the Institute's logo.

Woodhead books published in collaboration with The Textile Institute are offered to Textile Institute members at a substantial discount. These books, together with those published by The Textile Institute that are still in print, are offered on the Elsevier website at: http://store.elsevier.com/. Textile Institute books still in print are also available directly from the Institute's web site at: www.textileinstitutebooks.com.

A list of Woodhead books on textiles science and technology, most of which have been published in collaboration with the Textile Institute, can be found towards the end of the contents pages.

Related Titles

High Performance Textiles and Their Applications
(ISBN 978-1-84569-180-6)

Smart Textiles for Protection
(ISBN 978-0-85709-762-0)

Functional Textiles for Improved Performance, Protection and Health
(ISBN 978-1-84569-723-5)

Woodhead Publishing Series in Textiles:
Number 185

Performance Testing of Textiles

Methods, Technology and Applications

Edited by

Lijing Wang

The Textile Institute

ELSEVIER

AMSTERDAM • BOSTON • CAMBRIDGE • HEIDELBERG
LONDON • NEW YORK • OXFORD • PARIS • SAN DIEGO
SAN FRANCISCO • SINGAPORE • SYDNEY • TOKYO
Woodhead Publishing is an imprint of Elsevier

WP
WOODHEAD
PUBLISHING

Published by Woodhead Publishing in association with The Textile Institute
Woodhead Publishing is an imprint of Elsevier
The Officers' Mess Business Centre, Royston Road, Duxford, CB22 4QH, UK
50 Hampshire Street, 5th Floor, Cambridge, MA 02139, USA
The Boulevard, Langford Lane, Kidlington, OX5 1GB, UK

Library of Congress Cataloging-in-Publication Data
A catalog record for this book is available from the Library of Congress

British Library Cataloguing-in-Publication Data
A catalogue record for this book is available from the British Library

ISBN: 978-0-08-100570-5 (print)
ISBN: 978-0-08-100578-1 (online)

For information on all Woodhead publications
visit our website at https://www.elsevier.com/

Publisher: Matthew Deans
Acquisition Editor: David Jackson
Editorial Project Manager: Edward Payne
Production Project Manager: Omer Mukthar
Designer: Maria Ines Cruz

Contents

List of contributors

M. Azam Ali University of Otago, Dunedin, New Zealand

J. Beringer Hohenstein Institute for Textile Innovation GmbH, Boennigheim, Germany

A. Chiolerio Italian Institute of Technology, Torino, Italy

E. Classen Hohenstein Institute for Textile Innovation GmbH, Boennigheim, Germany

S. Geller Dresden University of Technology, Dresden, Germany

M. Gude Dresden University of Technology, Dresden, Germany

K. Holeczek Dresden University of Technology, Dresden, Germany

R.M. Laing University of Otago, Dunedin, New Zealand

S. Mandal University of Alberta, Edmonton, AB, Canada

N. Modler Dresden University of Technology, Dresden, Germany

S. Morlock Hohenstein Institute for Textile Innovation GmbH, Boennigheim, Germany

B.E. Niven University of Otago, Dunedin, New Zealand

R. Paul Hohenstein Institute for Textile Innovation GmbH, Boennigheim, Germany

X. Qiu RMIT University, Melbourne, VIC, Australia

A. Schmidt Hohenstein Institute for Textile Innovation GmbH, Boennigheim, Germany

A. Shavandi University of Otago, Dunedin, New Zealand

M.J. Smith University of Minho, Braga, Portugal

G. Song Iowa State University, Ames, IA, United States

M. Stoppa Italian Institute of Technology, Torino, Italy

M.D. Teli Institute of Chemical Technology, Mumbai, India

T. Tyczynski Dresden University of Technology, Dresden, Germany

H.L. Wainwright E-Textile Technology Consultant to fashion designers, Monessen, PA, USA

T. Weber Dresden University of Technology, Dresden, Germany

C.A. Wilson University of Otago, Dunedin, New Zealand

A. Winkler Dresden University of Technology, Dresden, Germany

Woodhead Publishing Series in Textiles

1 **Watson's textile design and colour Seventh edition**
Edited by Z. Grosicki
2 **Watson's advanced textile design**
Edited by Z. Grosicki
3 **Weaving Second edition**
P. R. Lord and M. H. Mohamed
4 **Handbook of textile fibres Volume 1: Natural fibres**
J. Gordon Cook
5 **Handbook of textile fibres Volume 2: Man-made fibres**
J. Gordon Cook
6 **Recycling textile and plastic waste**
Edited by A. R. Horrocks
7 **New fibers Second edition**
T. Hongu and G. O. Phillips
8 **Atlas of fibre fracture and damage to textiles Second edition**
J. W. S. Hearle, B. Lomas and W. D. Cooke
9 **Ecotextile '98**
Edited by A. R. Horrocks
10 **Physical testing of textiles**
B. P. Saville
11 **Geometric symmetry in patterns and tilings**
C. E. Horne
12 **Handbook of technical textiles**
Edited by A. R. Horrocks and S. C. Anand
13 **Textiles in automotive engineering**
W. Fung and J. M. Hardcastle
14 **Handbook of textile design**
J. Wilson
15 **High-performance fibres**
Edited by J. W. S. Hearle
16 **Knitting technology Third edition**
D. J. Spencer
17 **Medical textiles**
Edited by S. C. Anand
18 **Regenerated cellulose fibres**
Edited by C. Woodings
19 **Silk, mohair, cashmere and other luxury fibres**
Edited by R. R. Franck

20 **Smart fibres, fabrics and clothing**
Edited by X. M. Tao
21 **Yarn texturing technology**
J. W. S. Hearle, L. Hollick and D. K. Wilson
22 **Encyclopedia of textile finishing**
H-K. Rouette
23 **Coated and laminated textiles**
W. Fung
24 **Fancy yarns**
R. H. Gong and R. M. Wright
25 **Wool: Science and technology**
Edited by W. S. Simpson and G. Crawshaw
26 **Dictionary of textile finishing**
H-K. Rouette
27 **Environmental impact of textiles**
K. Slater
28 **Handbook of yarn production**
P. R. Lord
29 **Textile processing with enzymes**
Edited by A. Cavaco-Paulo and G. Gübitz
30 **The China and Hong Kong denim industry**
Y. Li, L. Yao and K. W. Yeung
31 **The World Trade Organization and international denim trading**
Y. Li, Y. Shen, L. Yao and E. Newton
32 **Chemical finishing of textiles**
W. D. Schindler and P. J. Hauser
33 **Clothing appearance and fit**
J. Fan, W. Yu and L. Hunter
34 **Handbook of fibre rope technology**
H. A. McKenna, J. W. S. Hearle and N. O'Hear
35 **Structure and mechanics of woven fabrics**
J. L. Hu
36 **Synthetic fibres: Nylon, polyester, acrylic, polyolefin**
Edited by J. E. McIntyre
37 **Woollen and worsted woven fabric design**
E. G. Gilligan
38 **Analytical electrochemistry in textiles**
P. Westbroek, G. Priniotakis and P. Kiekens
39 **Bast and other plant fibres**
R. R. Franck
40 **Chemical testing of textiles**
Edited by Q. Fan
41 **Design and manufacture of textile composites**
Edited by A. C. Long
42 **Effect of mechanical and physical properties on fabric hand**
Edited by H. M. Behery
43 **New millennium fibers**
T. Hongu, M. Takigami and G. O. Phillips
44 **Textiles for protection**
Edited by R. A. Scott

45 **Textiles in sport**
Edited by R. Shishoo
46 **Wearable electronics and photonics**
Edited by X. M. Tao
47 **Biodegradable and sustainable fibres**
Edited by R. S. Blackburn
48 **Medical textiles and biomaterials for healthcare**
Edited by S. C. Anand, M. Miraftab, S. Rajendran and J. F. Kennedy
49 **Total colour management in textiles**
Edited by J. Xin
50 **Recycling in textiles**
Edited by Y. Wang
51 **Clothing biosensory engineering**
Y. Li and A. S. W. Wong
52 **Biomechanical engineering of textiles and clothing**
Edited by Y. Li and D. X-Q. Dai
53 **Digital printing of textiles**
Edited by H. Ujiie
54 **Intelligent textiles and clothing**
Edited by H. R. Mattila
55 **Innovation and technology of women's intimate apparel**
W. Yu, J. Fan, S. C. Harlock and S. P. Ng
56 **Thermal and moisture transport in fibrous materials**
Edited by N. Pan and P. Gibson
57 **Geosynthetics in civil engineering**
Edited by R. W. Sarsby
58 **Handbook of nonwovens**
Edited by S. Russell
59 **Cotton: Science and technology**
Edited by S. Gordon and Y-L. Hsieh
60 **Ecotextiles**
Edited by M. Miraftab and A. R. Horrocks
61 **Composite forming technologies**
Edited by A. C. Long
62 **Plasma technology for textiles**
Edited by R. Shishoo
63 **Smart textiles for medicine and healthcare**
Edited by L. Van Langenhove
64 **Sizing in clothing**
Edited by S. Ashdown
65 **Shape memory polymers and textiles**
J. L. Hu
66 **Environmental aspects of textile dyeing**
Edited by R. Christie
67 **Nanofibers and nanotechnology in textiles**
Edited by P. Brown and K. Stevens
68 **Physical properties of textile fibres Fourth edition**
W. E. Morton and J. W. S. Hearle
69 **Advances in apparel production**
Edited by C. Fairhurst

70 **Advances in fire retardant materials**
Edited by A. R. Horrocks and D. Price
71 **Polyesters and polyamides**
Edited by B. L. Deopura, R. Alagirusamy, M. Joshi and B. S. Gupta
72 **Advances in wool technology**
Edited by N. A. G. Johnson and I. Russell
73 **Military textiles**
Edited by E. Wilusz
74 **3D fibrous assemblies: Properties, applications and modelling of three-dimensional textile structures**
J. L. Hu
75 **Medical textiles and biomaterials for healthcare**
Edited by S. C. Anand, J. F. Kennedy, M. Miraftab and S. Rajendran
76 **Fabric testing**
Edited by J. L. Hu
77 **Biologically inspired textiles**
Edited by A. Abbott and M. Ellison
78 **Friction in textile materials**
Edited by B. S. Gupta
79 **Textile advances in the automotive industry**
Edited by R. Shishoo
80 **Structure and mechanics of textile fibre assemblies**
Edited by P. Schwartz
81 **Engineering textiles: Integrating the design and manufacture of textile products**
Edited by Y. E. El-Mogahzy
82 **Polyolefin fibres: Industrial and medical applications**
Edited by S. C. O. Ugbolue
83 **Smart clothes and wearable technology**
Edited by J. McCann and D. Bryson
84 **Identification of textile fibres**
Edited by M. Houck
85 **Advanced textiles for wound care**
Edited by S. Rajendran
86 **Fatigue failure of textile fibres**
Edited by M. Miraftab
87 **Advances in carpet technology**
Edited by K. Goswami
88 **Handbook of textile fibre structure Volume 1 and Volume 2**
Edited by S. J. Eichhorn, J. W. S. Hearle, M. Jaffe and T. Kikutani
89 **Advances in knitting technology**
Edited by K-F. Au
90 **Smart textile coatings and laminates**
Edited by W. C. Smith
91 **Handbook of tensile properties of textile and technical fibres**
Edited by A. R. Bunsell
92 **Interior textiles: Design and developments**
Edited by T. Rowe
93 **Textiles for cold weather apparel**
Edited by J. T. Williams

94 **Modelling and predicting textile behaviour**
Edited by X. Chen
95 **Textiles, polymers and composites for buildings**
Edited by G. Pohl
96 **Engineering apparel fabrics and garments**
J. Fan and L. Hunter
97 **Surface modification of textiles**
Edited by Q. Wei
98 **Sustainable textiles**
Edited by R. S. Blackburn
99 **Advances in yarn spinning technology**
Edited by C. A. Lawrence
100 **Handbook of medical textiles**
Edited by V. T. Bartels
101 **Technical textile yarns**
Edited by R. Alagirusamy and A. Das
102 **Applications of nonwovens in technical textiles**
Edited by R. A. Chapman
103 **Colour measurement: Principles, advances and industrial applications**
Edited by M. L. Gulrajani
104 **Fibrous and composite materials for civil engineering applications**
Edited by R. Fangueiro
105 **New product development in textiles: Innovation and production**
Edited by L.Horne
106 **Improving comfort in clothing**
Edited by G. Song
107 **Advances in textile biotechnology**
Edited by V. A. Nierstrasz and A. Cavaco-Paulo
108 **Textiles for hygiene and infection control**
Edited by B. McCarthy
109 **Nanofunctional textiles**
Edited by Y. Li
110 **Joining textiles: Principles and applications**
Edited by I. Jones and G. Stylios
111 **Soft computing in textile engineering**
Edited by A. Majumdar
112 **Textile design**
Edited by A. Briggs-Goode and K. Townsend
113 **Biotextiles as medical implants**
Edited by M. W. King, B. S. Gupta and R. Guidoin
114 **Textile thermal bioengineering**
Edited by Y. Li
115 **Woven textile structure**
B. K. Behera and P. K. Hari
116 **Handbook of textile and industrial dyeing. Volume 1: Principles, processes and types of dyes**
Edited by M. Clark
117 **Handbook of textile and industrial dyeing. Volume 2: Applications of dyes**
Edited by M. Clark

118 **Handbook of natural fibres. Volume 1: Types, properties and factors affecting breeding and cultivation**
Edited by R. Kozłowski
119 **Handbook of natural fibres. Volume 2: Processing and applications**
Edited by R. Kozłowski
120 **Functional textiles for improved performance, protection and health**
Edited by N. Pan and G. Sun
121 **Computer technology for textiles and apparel**
Edited by J. L. Hu
122 **Advances in military textiles and personal equipment**
Edited by E. Sparks
123 **Specialist yarn and fabric structures**
Edited by R. H. Gong
124 **Handbook of sustainable textile production**
M. I. Tobler-Rohr
125 **Woven textiles: Principles, developments and applications**
Edited by K. Gandhi
126 **Textiles and fashion: Materials design and technology**
Edited by R. Sinclair
127 **Industrial cutting of textile materials**
I. Viļumsone-Nemes
128 **Colour design: Theories and applications**
Edited by J. Best
129 **False twist textured yarns**
C. Atkinson
130 **Modelling, simulation and control of the dyeing process**
R. Shamey and X. Zhao
131 **Process control in textile manufacturing**
Edited by A. Majumdar, A. Das, R. Alagirusamy and V. K. Kothari
132 **Understanding and improving the durability of textiles**
Edited by P. A. Annis
133 **Smart textiles for protection**
Edited by R. A. Chapman
134 **Functional nanofibers and applications**
Edited by Q. Wei
135 **The global textile and clothing industry: Technological advances and future challenges**
Edited by R. Shishoo
136 **Simulation in textile technology: Theory and applications**
Edited by D. Veit
137 **Pattern cutting for clothing using CAD: How to use Lectra Modaris pattern cutting software**
M. Stott
138 **Advances in the dyeing and finishing of technical textiles**
M. L. Gulrajani
139 **Multidisciplinary know-how for smart textiles developers**
Edited by T. Kirstein
140 **Handbook of fire resistant textiles**
Edited by F. Selcen Kilinc

141 **Handbook of footwear design and manufacture**
Edited by A. Luximon
142 **Textile-led design for the active ageing population**
Edited by J. McCann and D. Bryson
143 **Optimizing decision making in the apparel supply chain using artificial intelligence (AI): From production to retail**
Edited by W. K. Wong, Z. X. Guo and S. Y. S. Leung
144 **Mechanisms of flat weaving technology**
V. V. Choogin, P. Bandara and E. V. Chepelyuk
145 **Innovative jacquard textile design using digital technologies**
F. Ng and J. Zhou
146 **Advances in shape memory polymers**
J. L. Hu
147 **Design of clothing manufacturing processes: A systematic approach to planning, scheduling and control**
J. Gersak
148 **Anthropometry, apparel sizing and design**
D. Gupta and N. Zakaria
149 **Silk: Processing, properties and applications**
Edited by K. Murugesh Babu
150 **Advances in filament yarn spinning of textiles and polymers**
Edited by D. Zhang
151 **Designing apparel for consumers: The impact of body shape and size**
Edited by M.-E. Faust and S. Carrier
152 **Fashion supply chain management using radio frequency identification (RFID) technologies**
Edited by W. K. Wong and Z. X. Guo
153 **High performance textiles and their applications**
Edited by C. A. Lawrence
154 **Protective clothing: Managing thermal stress**
Edited by F. Wang and C. Gao
155 **Composite nonwoven materials**
Edited by D. Das and B. Pourdeyhimi
156 **Functional finishes for textiles: Improving comfort, performance and protection**
Edited by R. Paul
157 **Assessing the environmental impact of textiles and the clothing supply chain**
S. S. Muthu
158 **Braiding technology for textiles**
Y. Kyosev
159 **Principles of colour appearance and measurement**
Volume 1: Object appearance, colour perception and instrumental measurement
A. K. R. Choudhury
160 **Principles of colour appearance and measurement**
Volume 2: Visual measurement of colour, colour comparison and management
A. K. R. Choudhury
161 **Ink jet textile Printing**
C. Cie
162 **Textiles for Sportswear**
Edited by R. Shishoo

163 **Advances in Silk Science and Technology**
Edited by A. Basu
164 **Denim: Manufacture, Finishing and Applications**
Edited by R. Paul
165 **Fabric Structures in Architecture**
Edited by J. Ignasi de Llorens
166 **Electronic Textiles: Smart Fabrics and Wearable Technology**
Edited by T. Dias
167 **Advances in 3D Textiles**
Edited by X. Chen
168 **Garment Manufacturing Technology**
Edited by R. Nayak and R. Padhye
169 **Handbook of technical textiles Second edition Volume 1: Technical Textile Processes**
Edited by A. R. Horrocks and S. C. Anand
170 **Handbook of technical textiles Second edition Volume 2: Technical Applications**
Edited by A. R. Horrocks and S. C. Anand
171 **Sustainable Apparel**
Edited by R. S. Blackburn
172 **Handbook of Life Cycle Assessment (LCA) of Textiles and Clothing**
Edited by S. S. Muthu
173 **Advances in Smart Medical Textiles: Treatments and Health Monitoring**
Edited by L. van Langenhove
174 **Medical Textile Materials**
Y. Qin
175 **Geotextiles**
Edited by R. M. Koerner
176 **Active Coatings for Smart Textiles**
Edited by J. L. Hu
177 **Advances in Braiding Technology: Specialized Techniques and Applications**
Edited by Y. Kyosev
178 **Smart Textiles and Their Applications**
Edited by V. Koncar
179 **Information Systems for the Fashion and Apparel Industry**
Edited by T. M. J. Choi
180 **Antimicrobial Textiles**
G. Y. Sun
181 **Advances in Technical Nonwovens**
G. Kellie
182 **Advances in Women's Intimate Apparel Technology**
W. Yu
183 **Clothing for Children and Teenagers: Anthropometry, Sizing and Fit**
N. Zakaria
184 **Woven Terry Fabrics: Manufacturing & Quality Management**
J. P. Singh & S. K. Verma
185 **Performance Testing of Textiles: Methods, Technology and Applications**
Edited by L. Wang

Section A

Performance testing techniques, technologies and standards

Design and analysis in textile research

1

R.M. Laing, C.A. Wilson, B.E. Niven
University of Otago, Dunedin, New Zealand

1.1 Introduction

The objective of this chapter is to provide an overview of approaches to designing and analyzing investigations on textiles, which are broadly defined as "fibers, filaments and yarns, natural and manufactured, and products for which these are a principal raw material…threads, cords, ropes and braids; woven, knitted and nonwoven fabrics, lace, nets, and embroidery; hosiery, knitwear and made-up apparel; household textiles, soft furnishings and upholstery; carpets and other floor coverings; technical, industrial and engineering textiles, including geotextiles and medical textiles" (The Textile Institute, 2015). Methods include experiments, surveys, case studies, and historical research. General principles are common to all these, as are considerations of sampling, types of data, and procedures for analysis. Examples of investigations involving fabrics, garments, and bedding are given, with comment on design and analysis.

The Concise Oxford Dictionary defines research as "…the systematic investigation into and study of materials and sources in order to establish facts and reach new conclusions" (Concise Oxford English Dictionary, 2011). The two important words/phrases are *systematic* and *materials/sources*. Approaches differ, for example:

- experimentation, where one or more variables are manipulated, and effects of this manipulation on other variables are examined
- surveys, involving people or events using one of several different forms (eg, oral—face-to-face, telephone; postal; online)
- case studies, typically highly focused investigations of one topic, group, individual, or product
- elucidation and examination of history using original sources, which are typically varied.

1.2 Ethical issues

1.2.1 General

Ethical considerations in research (eg, human, cultural, types of questions deemed appropriate) apply in many countries, and some of these considerations apply irrespective of the country in which the investigation is planned and/or conducted. Codes of responsible practice or conduct are assumed (ie, the topic is worth investigating; methods used, including methods for analysis and reporting are appropriate; health and safety policies of the country, institution/organization are met; and where

Performance Testing of Textiles. http://dx.doi.org/10.1016/B978-0-08-100570-5.00002-5

animals are involved in testing, practical aspects comply with appropriate regulatory organizations).

1.2.2 Ethical practices in research involving humans

Research involving human participants is typically governed by policies on ethical practices underpinned by a number of principles, eg, merit of the research, informed consent of participants, issues related to vulnerable subjects, privacy of participants, minimization of harm, limitation of deception, avoidance of conflicts of interest, cultural and social sensitivity, and publication of results. Ethics committees routinely require that consideration be given to a range of factors, and require prior decisions on:

- use of personal information collected or used;
- taking or handling any form of tissue or fluid samples from humans or cadavers;
- any form of physical or psychological stress;
- situations which place the safety of participants or researchers at risk;
- administration of food, fluids, or drugs to any participant;
- a potential conflict between the applicant's activities as a researcher, clinician, or teacher and their interest as a professional or private individual;
- any form of deception of the participant; and
- access to and storage of data and eventual disposal (eg, University of Otago, 2015a).

1.2.3 Cultural issues

Cultural issues may need to be taken into account prior to and during research planning and conduct (eg, indigenous issues in Canada, Australia, and New Zealand). In New Zealand, research "purchasers" and "providers" (eg, Crown Research Institutes, universities) all acknowledge the Treaty of Waitangi, New Zealand's founding document (New Zealand History, 2015). Further, the University of Otago in its Ethics Policy requires that research complies with this Treaty and that consultation is appropriate (University of Otago, 2015b).

1.2.4 Intellectual property

Who "owns" any particular idea, written material, or knowledge about textiles (a test method, a design) has been the subject of discussion at international fora (eg, meetings of the World Trade Organization, Organization for Economic Co-operation Development—OECD), and at national organizations and institutions. The issues are relevant to research in general and to the textile trade (eg, copying of textile and other designs). There are two aspects to managing intellectual property which impinge on researchers (1) use of copyright material, such as text, figures, tables by one other than the originator of that material; and (2) establishing ownership of original work. Copyright is relevant in reviewing literature, for paper submission to a refereed journal, or a report for a company. Ideas and concepts are to be correctly referenced, attributing them to their origin. As to intellectual property, many organizations take action with respect to their rights as owners of patentable discoveries which staff may develop in the course of their employment. Mutual benefit of students and staff is typical of universities. In cases where the research has been externally funded, decisions on ownership of the intellectual property is normally made and agreed on as part of the contract. However, much research is funded from public sources, and there is agreement that research data from public funding be made widely available (ie, OECD Declaration; Maslen, 2013).

1.3 Design and analysis: General considerations

1.3.1 Design

Sound investigative design is a basic requirement, irrespective of the type of investigation. Variability is inherent in all textile research and needs to be taken into account at the planning stage. Fabrics are variable (batch to batch, across one fabric piece), operators and participants are variable (efficacy of the operator may diminish over the day or week; individual participants differ, hence, many investigations are based on percentage change from baseline data), laboratories in which an investigation is conducted are likely to differ even when the same test is carried out. Accounting for these issues is achievable through representative sampling, an appropriate number of replicates/interviewees, randomization, and inclusion of controls (eg, before and after monitoring effects of operators, ambient conditions).

Randomization is to ensure bias does not exist in the overall sample, and that appropriate statistical analysis can be carried out. Replication (ie, the repetition of samples in each treatment, in an experiment) is planned and implemented to reduce standard error (Gacula and Singh, 2009). Also included in the planning phase is the planning for statistical analysis required for interpretation of data (Leedy and Omrod, 2016). When there are several sources of variation in a design (eg, nested analysis of variance), the appropriate variance estimates relevant to each test need to be used to ensure correct analysis and results are obtained. In chemistry, for example, triplicate samples cannot usually be treated as replicates in an experiment since they are designed to measure instrumental variance within a sample, not the variability between samples. Planning for these steps to be taken is required to limit error, and increase validity (Choi, 2008). With respect to surveys, a common flaw is failure to specify and conform to the definitions and objectives. Errors may be attributable to characteristics of the method itself (eg, respondent attrition in longitudinal studies), or arise from question format, order and/or content, or researcher effects (eg, poor measurement or questioning) (Sudman et al., 1977).

1.3.2 Analysis and significance

Data collected in experiments, surveys, case studies, and historical investigations may be qualitative or quantitative, each data form requiring consideration and selection of potential analysis procedures. Data differ in quality, and the range of statistical tests which are appropriate needs to be determined prior to data collection (Tables 1.1 and 1.2).

Table 1.1 Measurement scales and data classification

Measurement scale	Description
Nominal or categorical	When numbers or other symbols are used to classify an object, person or characteristic, measurement is at its weakest. Also known as classificatory scale, eg, classifying items
Ordinal or rank	Where objects in one category are not only different from those in other categories but also stand in some kind of relation to them, eg, grades in assessment A > B > C
Interval	This scale has the characteristics of an ordinal scale, and the distance or difference between two numbers has meaning, eg, temperature
Ratio	This scale has the characteristics of an interval scale, and has a true zero point as its origin, eg, weight
Data classification	
Parametric	A parametric statistical test specifies certain conditions about the distribution of the population from which it is drawn, eg, normality. The relevance of results depends on these assumptions being met. Measurements must be at least in an interval scale
Nonparametric	A nonparametric statistical test specifies very general conditions regarding the distribution of the population from which it is drawn. Certain assumptions are made, eg, that observations are independent, but they are weaker than those required for a parametric test. Measurements are generally nominal or ordinal
Sample relationships	
Related	When the same subject is tested under different conditions, samples from that subject are considered related
Unrelated/ independent	Two samples are regarded as independent if there is no basis for grouping or linking the values

Based on Siegel and Castellan (1988).

Not addressing analysis issues as an integral part of the research design may yield data which require recoding, manipulation, and cross-checking: worse, analysis may prove to be intractable.

The choice of analysis is influenced by the type of data and its distribution. Survey data, such as that collected during anthropometric studies (typically interval data), may be analyzed using parametric statistical techniques, while questionnaires (typically opinion-based), are more appropriately tested using nonparametric tests (nominal or ordinal data). In the case of questionnaires, the wording of questions, the organization of response categories, the method to categorize or code, and the format to collect the data all may influence the analysis method which can be used.

Table 1.2 Selecting an appropriate statistical test for comparing samples

Type of data	Selected examples	
a. Measures of association between two variables		
Interval	Pearson correlation	
Ordinal	Spearman rho	
	Kendall tau-a, tau-b, tau-c	
Nominal/categorical	Phi coefficient	
	Cramer V	
b. Comparing samples to determine whether they are significantly different		
	Independent	*Related*
Two samples		
Interval	Independent sample *t*-test	Paired sample *t*-test
Ordinal	Mann–Whitney *U*-test	Wilcoxon
		Sign
Nominal/categorical	Chi-square	McNemar
Three or more samples	Univariate ANOVA[a]	Repeated measures ANOVA
Interval	Mixed-model	Mixed-model
Ordinal	Kruskal–Wallis k-sample	Friedman
Nominal/categorical	Chi-square	Cochran's Q (dichotomous nominal data only)

[a] Analysis of variance.
Adapted from Siegel and Castellan (1988).

Preliminary analysis of data is a requirement. This involves checking for anomalies, outliers (consider whether these are errors or real data points), determining the frequency of response, comparing group responses in a data set, possibly exploring the size of the effect and if necessary, modifying the sample size for use in designing future experiments.

When data cannot be numerically quantified or are ordinal, nonparametric statistical tests are used to determine significance of differences. Tests might include two-sample tests (Wilcoxon signed rank, Mann–Whitney rank sum tests), nonparametric analysis of variance (Kruskal–Wallis one-factor analysis, Friedman test for related data), or rank correlation (eg, Spearman). One disadvantage of nonparametric tests is they do not use as much information as parametric tests: they make less stringent demands on the data and therefore it becomes more difficult to make qualitative statements about actual differences between populations, ie, they are less efficient (Harraway, 1993). Nevertheless, their use is appropriate with some types of data (Tables 1.1 and 1.2). Parametric tests are used when data can be quantified, and when assumptions about the distribution of the sample (ie, normality) and data (ie, interval or ratio) can be made. For example, *t*-tests can be used for simple comparisons of means, while analysis of variance (using a single measurement variable) can be used to examine relationships between one or more factors thought to influence the experiment and the data. Correlation would be determined using Pearson's procedure. Measurements repeated on the same specimen or participant might be analyzed using repeated measures analysis of variance. The number of factors used depends on the complexity of the experiment (Harraway, 1993). Research on textiles and clothing can involve quite complex situations with a number of measurement variables which are analyzed together using multivariate statistical methods. Multiple analysis of variance, principal component analysis, factor analysis, and cluster analysis which are all used in textiles research, are described by Manly (2005).

With respect to interpretation of results, variability is one of the key considerations: how variable were the materials being examined (ie, how large is the effect being investigated), was there evidence of the operator changing competence over time (ie, the operator became more familiar with the test instrument or the set of questions; or conversely became tired), did differences in day-to-day ambient conditions appear to have any effect?

Most research begins with a hunch, a proposition, or a theory, which may be expressed formally as a hypothesis, typically in null form (ie, no detectable effect). Statistical tests are then used to determine whether the hypotheses should be accepted or rejected. Whether rejection is appropriate depends on the test statistic used, the magnitude of differences among treatments, the variability among sample values, and the level of significance which has previously been agreed to by the researchers. Levels of significance of 0.05 or 0.01 are often considered appropriate in textiles: this means there is a 5% or 1% chance of rejecting the null hypothesis based on the statistics when it is the correct hypothesis (Type I error). In other situations, a 0.001 level of significance is required, and in these circumstances greater attention needs to be given to the test design and method, the cost of carrying out the study, and consequences of a wrong conclusion. Note also that a test can fail to reject the null hypothesis when it is not the correct one (Type II error). Inherent variability of the particular system being examined determines what level of significance is possible and whether differences in the data can be detected.

Methods for investigations on textiles and garment systems, particularly procedures for data collection, are diverse. These may include application of an existing standard test method (eg, ISO, EN, BS, AATCC), a method published in a refereed research journal, modification of an existing method, or development of a new approach. In general, unless there is good reason for not doing so (ie, a desire to better simulate a real event), use of an existing method is preferred. Many physical tests of textiles (fibers, yarns, fabrics, garments, bedding, for example), require compliance with controlled or standard conditions such as those set out in ISO 139: 2005 Textiles—standard atmospheres for conditioning and testing $20 \pm 2°C$, $65 \pm 4\%$ R.H. (International Organization for Standardization, 2005). This standard specifies prior conditioning of the textile for at least 24 h, with tests conducted under these conditions. There is evidence that precleaning fabrics stabilizes dimensions and ensures finishing agents have been removed, and thus, precleaning of test fabrics is considered good practice (Gore et al., 2006) as variability among specimens, due to factors other than those of interest, is reduced.

1.4 Approaches to research and their application

1.4.1 The experiment

Consider a simple, functional experiment: suppose we want to determine if seams constructed using a single jersey fabric in two different stitch types (eg, stitch type ISO 301 and stitch type ISO 401; ie, cause, two "treatments"), differ in their extensibility (ie, effect). We first prepare the fabric specimens, ensuring that in cutting, different wales and courses are used (European Committee for Standardization, 1999), and we allocate an equal number of samples of fabric at random to each stitch treatment. Stitch and seam parameters (eg, stitch length, yarn, seam length) are matched. The seamed specimens are randomized in order of testing as all specimens would not be tested in one day. To check whether or not the conditions of the test (operator, instrument, ambient conditions) change over time, we include a control at the beginning and end of each day, and possibly part way through the day. We anticipate the controls will not differ significantly from each other, and that if they do, we need to consider carefully how to proceed with any analysis of that experiment. Breaking strength would normally be analyzed using an unpaired t-test, the equivalent to a one-way analysis of variance with two treatments. As is the case for all statistical tests, assumptions must be verified before proceeding to form conclusions. In this case the assumptions are independence of data points, normality of the distributions within treatments, and usually, equality of treatment variance. Parameters for the test itself need to be the subject of preliminary testing, selecting practical size intervals over the range for the effect of interest.

Functional experimental designs involve use of independent variables controlled by the experimenter, whereas factorial designs are those in which independent variables cannot necessarily be directly controlled (Leedy and Omrod, 2016; Davies, 1979). Factorial designs have their origin in agriculture, where variables (eg, soil type, moisture, angle of the sun) can rarely be managed. This is also true of much research on textiles: for example, in a comparison of effects of different fibers on fabric properties, are the fiber properties (eg, length, diameter) "the same"; have yarns been manufactured such that the structure is "the same"; are fabric structures (eg, mass-per-unit area, thickness, permeability to air) "the same"; have the fabrics been finished in the same way? In general, the simplest design for answering the research question should be selected (Cochran and Cox, 1992).

1.4.1.1 Example—The factorial experiment

We may be interested in determining whether the type of fiber or the type of fabric structure dominates the tendency for a next-to-skin fabric to retain body odor. The experiment needs to be designed. Perhaps there is more than one experiment, and the various experiments need to be considered together. Indeed, McQueen's questions involved several types of investigation—human sensory perception (McQueen et al., 2007a), instrumental determination of volatiles (McQueen et al., 2008), and links between odor volatiles and microorganisms resident on the human skin (McQueen et al., 2007b).

McQueen approached these questions with a factorial design (McQueen et al., 2007a,b). Three fiber types (wool, cotton, and polyester), and three fabric structures (single jersey, 1×1 rib, interlock), typical of next-to-skin garments, were selected. A particular challenge in dealing with investigations such as this is to ensure to the extent practicable, that all manufacturing parameters are controlled (ie, yarn and its supply, the knitting machine and its operation, application of fabric finishing processes) and that resultant fabric properties (ie, mass-per-unit area, thickness) are also well matched. Without this care, we are left querying whether any differences observed relate to either of the key parameters in question (ie, to the fiber type or the fabric structure).

With respect to human sensory perception there is a need to train those involved in detecting the odor volatiles to ensure variability among participants is minimized. In McQueen's work (McQueen et al., 2007b), the paired-comparisons procedure was used (ie, is one of the pair more odorous than the other?), along with a line scale noting the perception of intensity. The paired-comparisons are likely to give better discrimination between samples, but are more time intensive compared to having the participants simply rank order the specimens presented. Since odor is related to volatiles emitting from the textile, some practical aspects of the investigation involved collecting axillary odor on fabrics, and storage of these fabrics for evaluation of aromatic qualities. The quad design was used: this design allows for a balanced set of pairs from each participant, when some aspect of the experiment changes in a systematic way (eg, over time, carpet; Miller, 2002), or that the time involved becomes too great because of the number of samples and subjects (eg, scoured silk; Kim et al., 2005). McQueen's experiment was analyzed using the statistic peculiar to this design, TSTAT for rank data (distributed as χ^2) and t-distributions for significance of difference among rank sums of fabrics compared. The performance of the panel was evaluated using one-way analysis of variance to determine the ability of each assessor to discriminate among samples, and a Pearson coefficient used to determine the correlation between the score from each assessor and the mean scores from the panel as a whole. McQueen correctly points out that while the trend in analyzing volatiles is toward instrumental methods, it is humans who provide real life perceptions, as well as providing the basis for the complex interactions among axillary microflora, physiological secretions, and mechanical action of the underarm.

Linking sensory perception of odor intensity to the presence of microorganisms retained on a fabric after exposure to the axillae and stored for three time periods (1, 7, 28 days), was also investigated by McQueen et al. (2007a). The question here is this: we believe that human odor is formed by effects of bacteria on body secretions, and that wool and polyester fabrics in particular seem to behave differently in this regard, the wool not being so odoriforous, but what is the evidence for this? Small specimens sampled randomly from a main fabric patch were used during each of the three time periods. A control sample was always presented as the first sample and the order of presentation of other samples was based on the 18-treatment Williams' design to reduce possible effects of order (MacFie et al., 1989). This type of design is balanced for first order carry-over effects, and requires relatively few subjects to achieve this. Each ordered pair of treatments occurs equally often. Each of 13 participants had shown good reliability and discrimination in sensory perception. The microbiological analysis involved splitting specimens into groups to reduce the effect of unevenly-distributed population densities of the bacteria on the specimens, and providing three types of culture media. Viable counts were expressed as colony forming units per milliliter. Fabrics were handled in aseptic conditions throughout. Data in the sensory part of the investigation were analyzed by repeated measures analysis of variance with time period as the repeated measures factor. In this example, analysis of variance was performed on $\log_{10}$ transformed bacterial counts to determine whether any difference existed. The transformation was to ensure assumptions of normality, a requirement for analysis of variance, were met, and although other transformations were trialled, $\log_{10}$ was the most appropriate. Tukey's Honest Significant Difference tests were then used to identify specific differences in odor intensity and in microbiological data. Whether parametric or nonparametric procedures are most appropriate for analysis of sensory data continues to be a point of debate, and practices on this differ.

Having demonstrated that odor of high intensity emanated from polyester fabrics and that this continued to emanate from polyester fabrics even after 28 days, McQueen sought to detect possible compounds in the headspace above the wool, cotton, and polyester fabrics using the instrumental technique of proton-transfer-reaction spectrometry (McQueen et al., 2008). Malodors continued to increase over 7 days with the polyester fabrics, but not with either the wool or the cotton fabrics, with the intensity of axillary odor seemingly inversely related to fiber hygroscopicity.

Mixed models provide another useful tool for analyzing physical relationships between humans and textiles. Consider the question as to whether or not compression garments used in sport have a desirable physiological effect. Detection of such an effect can be determined only through a human trial. MacRae examined several of these, including thermal effects of upper- and lower-compression garments on twelve recreationally trained male cyclists (a sample whose characteristics were similar): was the effect simply one of the surface area of the body covered, hence ensuring a higher skin temperature, or were there other benefits (MacRae et al., 2011)? Data were collected during rest and fixed-load exercise, and included cardiovascular and thermoregulatory responses (both ratio data), and sensory ratings (ordinal data). Important in these types of investigations is collection of sufficient baseline data for comparison against exercise data. Should gear failure occur, missing data may be estimated (if only a few points) or in extreme cases, data from one participant may be unusable. This means that it is important to include two or three participants more than the design requires to accommodate this eventuality. Data from the rest and exercise periods were analyzed separately. A linear mixed model with an unstructured covariance matrix for the repeated measure was used to test for differences among garment types and, where applicable, time periods for subgarment pressures, cardiovascular variables, core and skin temperatures, forearm perfusion, whole-body sweat rate, next-to-skin vapor pressures, and time-trial duration, mean power, and cadence. MacRae and coworkers did trial other covariance structures (including diagonal, compound symmetry, compound symmetry heterogeneous), but some were considered less appropriate and others failed to fit (MacRae et al., 2011). Failure to fit is often due to insufficient data in a treatment, having too simple a covariance structure, or the use of too many parameters relative to the size of the data set. The simplest covariance structure of the data is usually chosen. The effect of garment type on the psychophysical sensory ratings, and rank data, were analyzed using the Friedman nonparametric analysis of variance (Siegel and Castellan, 1988).

1.4.2 The survey

Surveys involve collection of information through a series of questions (eg, questionnaires), by direct measurement of physical properties (eg, anthropometric surveys), or some other process. Depending on the sample selected, the profiles or statistics derived may be used to describe a broader population: therefore, sample development and selection involve critical decisions and forward thinking. Two other differences in types of surveys relate to time and to mode of completion. Time may be either cross-sectional, involving a sample at one time point (typically a large sample, with emphasis on securing a random sample to ensure findings will be applicable beyond the group studied), or longitudinal, involving data collection on individuals or groups over time (eg, body sizes changing over time; Noppa et al., 1980). Completion may be self-completion (eg, mailed questionnaires, computer-based surveys), or administered interviews (eg, face-to-face, telephone).

Factors likely to influence the choice of method include the nature of the questions, likely response rate, resources, and the time available for obtaining results. Formulation of the objective and clarity in the intended use of the survey findings are essential steps before developing the survey design and analysis. Methods of sample selection, data entry, transcription, storage, and analysis also need to be determined prior to data collection. The language of questions needs to be pitched to the user group, ie, unambiguous, in a logical order, and expressed in a neutral manner. A useful step is to trial the questions on a small group of respondents before the survey is conducted—Is the sequence appropriate, are the possible responses from which to select clear, how long does the set of questions take to complete?

1.4.2.1 Examples—The survey

Body dimensions of groups of people required to wear personal protective clothing systems or groups which form a market segment are of interest to manufacturers, employers, and distributors. The body dimensions of various groups and subgroups of people have been surveyed over many years using direct measurement by Pheasant and Haslegrave (2005). Data are summarized typically in table form (mean, standard deviation, coefficient of variation; often accompanied by percentiles, 5th, 50th, 95th). However, for use in development of textile products for humans (ie, product sizing), this type of information requires further analysis.

The first example is an investigation for the New Zealand Fire Service. We were asked to develop a suitable sizing system for the active firefighting personnel (Laing et al., 1999). The sample, about 7.5% of the workforce (n=750), was selected from two regions, with random selection by fire appliance (four participants per appliance), watch (four watches), and employment category (permanent, volunteer). Although an earlier investigation had shown no significant difference in body dimensions of males from the northern and southern regions of New Zealand (Wilson et al., 1990), we opted to sample from two sites to encourage acceptance of findings. To ensure measurers were measuring in a reproducible way, we re-measured approximately 10% of the sample. In analyzing the data, whether data from the various groups (two geographic regions; contract type) could be pooled, was established by checking for significance of difference in 12 measurements. Pooling was determined as being appropriate. These steps demonstrate care in ensuring (1) the sample was representative of the role in the service, the geographic region, and the service as a whole, and (2) the measuring process itself was reproducible.

The next step was to determine sizing criteria. All measurements of a body site related to a product/covering were identified. For a head covering—head depth, head girth, head breadth, and face length: for a foot/lower limb cover—calf girth, ankle girth, foot breadth, foot girth, tibiale height, and foot length. The measurements accounting for the most difference were identified using factor analysis based on principal component analysis (Manly, 2005), with the percent of variance explained by the various measurements noted (Table 1.3) (Laing et al., 1999). That process enabled us to clarify probable control variables for each product (and compare variables considered traditional and those identified through statistical procedures).

The final step was to develop size groups. This required some knowledge of the commercial implications of more or fewer sizes being produced, and the fabric/material properties (eg, elastic or non-stretch). Each body section was divided into size groups using the K-means clustering technique (Hair et al., 2010). Outliers were identified in two ways: first by examining the minima and maxima for each variable, and second through Mahalanobis distances in multivariate space (Manly, 2005). Outliers in the data sets were included in summary data, but were excluded from the principal component analysis and K-means clustering because protective items for atypical body shapes would be made-to-measure.

In the second example, retrospective and prospective surveys are considered. Injury to the pretibial region of people aged 50 years and older is common and presents ongoing and often debilitating effects for sufferers and their families. The questions were: what are the circumstances surrounding these types of injury events; do the people involved exhibit similar characteristics such as age, ethnic origin, sex; would some sort of covering over the lower leg provide a means of reducing the incidence of injury or its severity; what do these people typically wear? A series of surveys and experiments were carried out to provide answers to these questions: (1) a retrospective study was conducted using data on patients with this type of injury who had been discharged from any New Zealand public hospital over a specified time period (Laing et al., 2002); (2) a prospective study was undertaken to identify what was worn by the patients and whether what was worn could be linked to the severity of the injury sustained (Laing et al., 2008); and (3) a laboratory test was developed to determine the potential protection against impact provided by typical apparel fabrics and combinations (Laing et al., 2008). The prospective survey is of interest here: a record of events surrounding a pretibial injury was made of consecutive injured patients presenting to the Wellington Regional Plastic, Maxillofacial and Burns Unit at Hutt Hospital, Wellington, New Zealand. The sample consisted of all persons aged 50 years and

Table 1.3 Principal components for body sections (a) and size charts (b)

(a) Principal components for body sections (decreasing order by coefficient size)

Head—total variance explained by three derived factors 95.9% (n= 682)

Upper head	Head breadth	Face length
54.5%	21.3%	20.1%
(Head depth)		
(Head girth)		

Hand—total variance explained by three derived factors 88.5% (n= 683)

Hand length	Hand breadth and girth	Lower upper limb girths
56.9%	19.1%	12.5%
(Palm length)		
(Hand length)		

Foot—total variance explained by four derived factors 88.4% (n= 681)

Lower limb girths	Foot dimensions	Lower limb lengths	Ankle height
49.4%	18.7%	11.2%	9.1%
(Calf girth)	(Foot breadth)	(Tibiale height)	(Ankle height)
(Ankle girth)	(Foot girth)	(Foot length)	

Lower torso—total variance explained by four derived factors 86.0% (n= 675)

Lower lengths	Lower trunk girths	Lower limb girths	Ankle height
50.3%	26.4%	5.1%	4.2%
(Crotch height)			
(Waist height)			
(Tochanterion height)			
(Knee height)			
(Popliteal height)			
(Stature)			
(Tibiale height)			
(Buttock-knee length)			
(Buttock-popliteal length)			

Upper torso—total variance explained by five derived factors 75.8% (n= 672)

Upper torso girths	Upper limb lengths	Torso lengths	Shoulder breadth	Hand/wrist girth
39.8%	18.4%	7.0%	5.6%	5.0%
(Chest girth)				
(Waist girth)				
(Anterior-posterior chest depth)				
(Transverse chest breadth)				
(Weight)				
(Arm girth relaxed)				

Total body cover—total variance explained by four derived factors 72.2% (n= 654) (excluding head, hands and feet)

Lengths/heights	Weight/girths	Torso proportions	Shoulder breadths
42.5%	20.5%	5.1%	4.0%
(Crotch height)			
(Spinale height)			
(Knee height)			
(Waist height)			
(Trochanterion height)			
(Popliteal height)			
(Tibiale height)			
(Shoulder-elbow-wrist length)			
etc. (12 others including stature)			

Table 1.3. **Continued.**

(b) Typical size chart. Foot coverings (based on foot length; mm body dimensions (unless specified)

Number of existing footwear sizes		11							
Solution		8 sizes (but recommend one further size at each end to accommodate all employees)							
Tolerance		10							
Centre of size		235	245	255	265	275	285	295	305
Range of size		230–240	240–250	250–260	260–270	270–280	280–290	290–300	300–310
Approximate % of sample fitted*		< 1	5	18	33	31	17	5	< 1
% of sample fitting more than 1 size		10							
% not fitted (outliers)		1							
Body dimensions									
Ankle girth	$\bar{X}$	222	218	224	228	231	238	240	236
	5%	–	198	207	211	215	220	221	–
	95%	–	243	242	246	250	260	261	–
Ankle height	$\bar{X}$	127	130	132	135	139	142	144	147
	5%	–	109	116	119	123	127	127	–
	95%	–	148	151	153	156	158	158	–
Foot girth	$\bar{X}$	258	254	261	266	271	277	280	281
	5%	–	234	243	249	254	263	264	–
	95%	–	273	281	284	289	299	301	–
Foot breadth	$\bar{X}$	104	100	104	106	108	111	112	111
	5%	–	92	93	97	98	103	101	–
	95%	–	109	113	114	117	120	121	–
Tibiale height	$\bar{X}$	412	421	430	442	454	463	481	508
	5%	–	382	395	409	422	432	443	–
	95%	–	467	461	469	489	496	518	–

*Rounding errors mean that the total is not always 100%.
(Laing et al., 1999)

older presenting to the unit with a pretibial injury until the sample size required for the specified power level was obtained ($n-75$). So in this example, the time required to conduct the survey could not be established with any confidence in advance. Rather, the investigator team continued to interview until the required number of participants was reached. Patients seen at Hutt Hospital Emergency Department who were deemed to not require surgery were not included in the study (an unknown number in this category), hence those in the survey were likely to have been among the more severely injured. These were the patients for whom possible interventions were to be focused.

1.4.3 The case study

A case study is an in-depth investigation of one or more subjects (cases). It may be exploratory (ie, aimed at determining feasibility, defining hypotheses or questions intended for subsequent study), descriptive (ie, providing a complete description of a particular case), or explanatory (ie, determining which causes produce which effects). Many techniques may be used to collect the data in a case study, including interviews (open-ended, structured, unstructured), and direct observation (of individuals, products or situations, under controlled or natural environments). Reliability of observational data can be strengthened with photographic or videotaped evidence, with or without time-sampling.

1.4.3.1 Example—The case study

One step in developing and refining a product, in this case, a cooling vest is to undertake a field trial: the question related to how the product performs under the conditions in which it was designed to be used. Heat stress is a problem in many workplaces, cause by any combination of ambient conditions of the workplace, and physiological effects of the activity undertaken. The cooling vest was intended to be used in a plant manufacturing domestic kitchen ovens (high ambient temperature, minimal air flow), and during road construction on public roads (in regions of high temperature and intense sun) (Laing et al., 2006). Data were primarily observational and user responses to questions, nominal and ordinal data respectively. This case study was successful in that practical difficulties in using the cooling vest were identified.

1.4.4 The investigation of historical evidence

Historical research refers to past events and the way these are represented and in some cases also explained. It is typically multidisciplinary. This research method involves posing and answering one or more questions, and using primary (original) and secondary source material. The sources need to be scrutinized—are they authentic, are they complete, what was the context in which the material existed? In many areas of clothing and textiles, items may have been partially or fully destroyed, and record-keeping may have been minimal or nonexistent. Survival of "evidence" may reflect effects of class, ethnicity, location, and/or storage.

While written sources are common in historical studies generally (witness and participant accounts; printed documents; government, organizational, and private papers), in terms of clothing and textiles, paintings, photographs, film, artifacts and artifact remnants, archeological imprints, all have the potential to provide rich sources of information. Irrespective of form, critical review is required to determine relevance and meaning.

1.4.4.1 Example—Historical evidence

In her study of settler dress in New Zealand, many sources of evidence were examined by Malthus (eg, diaries, newspapers, published historical writings, and collections of garments in museums). A collection of garments for women believed to be from the period 1828 to 1914 was described fully (eg, dimensions, fabric structure, and design features). These data include both nominal and interval types. Using the design and other features, the time period in which the garment had been used was estimated through two-way (ie, dress identifiers versus dress features) block cluster analysis (Malthus, 1996; Hartigan, 1972). This analytical procedure groups items which have properties in common. Interval data were split into two categories using the median, but this was found to not be useful in the final model. In the example from Malthus, a time-period 1860–1900, which had not been identified from the more common visual examination of artifacts, photographs, and newspapers and which consisted of 24 dresses, was revealed.

1.4.5 Modeling

Various forms of models are used in research, either as a starting point or modified/developed as part of the research itself. Types of models include (1) a theoretical model, based on a theory which may exist prior to undertaking the research (ie, the theory of heat transfer) or be developed or modified during the research (eg, mechanical properties of stitches in a weft knit textile; Webster et al., 1998); (2) a physical model (eg, layers of fabric in a series of textile layers simulating infant bedding assembly) (Wilson et al., 1999); or (3) a statistical model (eg., to elucidate dominant factors and/or relationships among variables in the experiment).

1.4.5.1 Example—Modeling

The example of infant bedding, with particular attention to thermal resistance of the assembly, is of wide interest insofar as "overheating" was one of several possible explanations advanced during the early 1990s for unexplained sudden infant death (Wilson et al., 1994). Thermal resistance of textiles results primarily from thickness. So could thermal resistance of bedding not be the sum of the resistance of the various layers? Even ignoring a compressive effect of multiple bedding layers on total thickness (and thus thermal resistance), using a physical model, Wilson was able to improve our understanding of thermal resistance in this application (Fig. 1.1).

This model involved identifying variables which contributed to thermal resistance of bedding as it is used (ie, in a three-dimensional configuration). In earlier investigations, thickness of flat bedding items had been used to predict thermal resistance, and then as an indirect indicator of risk of an infant overheating. Development of the model necessitated design of a series of related experiments: (1) variables affecting the arrangement of infant bedding during use were identified (sleep position, tucking arrangement, specific bedding items used, and in what combination (sheet, blanket(s), duvet; Wilson et al., 1994); (2) the cumulative thickness of each bedding layer, including the underlying air, at approximately the shoulder was measured across the width of a bed, for all bedding, sleep and tucking combinations (Wilson et al., 1999) allowing the arrangement of bedding and air in use to be mapped (Fig. 1.1); (3) the effect of bedding and air on thermal and water vapor resistance of the assembly was determined, with fabrics arranged flat but with layers separated such that the dimension of each air space matched specific sleep positions, tucking arrangements and bedding combinations (Wilson et al., 1999, 2000), and (4) a theoretical model for predicting thermal resistance of bedding was developed (Wilson et al., 2002) and validated (Wilson et al., 2004). Each investigative step in the series involved the same combination of variables, but in different contexts, thereby allowing the effect of arrangement to be scrutinized.

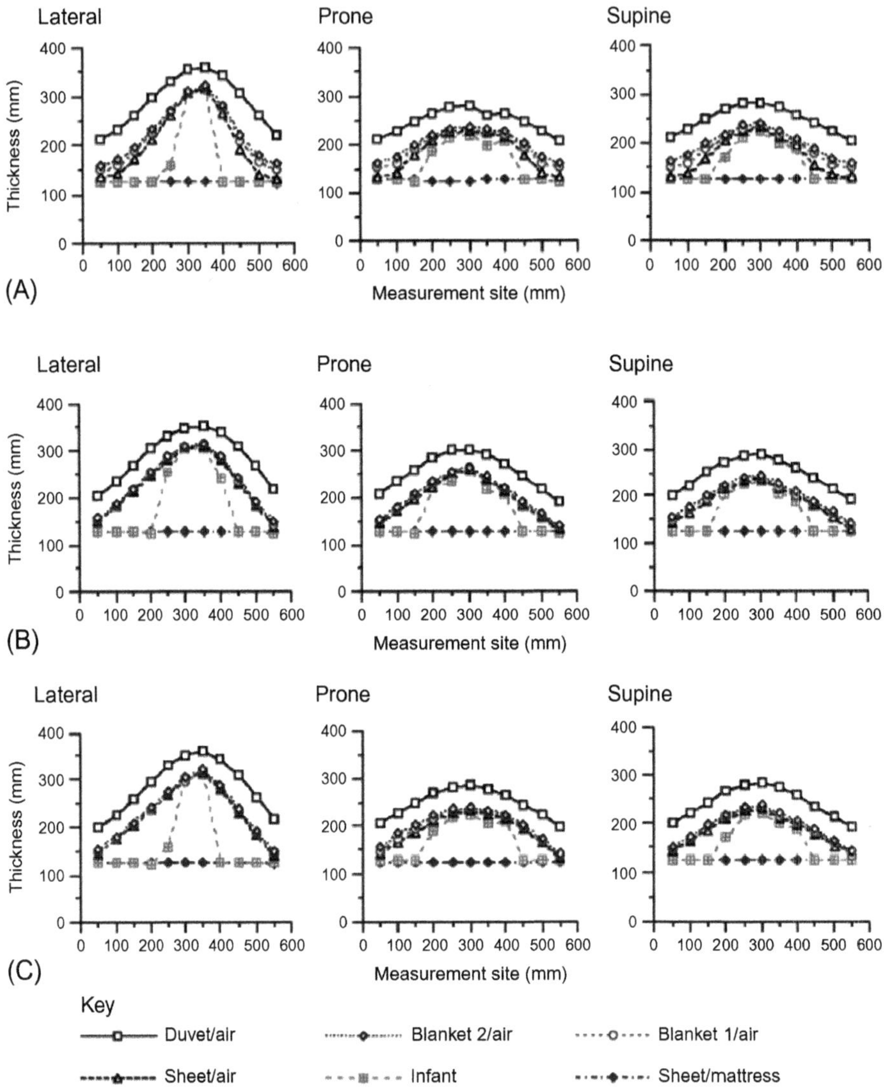

Fig. 1.1 Profile of sheet, 2 blankets and a duvet (and underlying air space) formed when arranged over lateral, prone, or supine baby with bedding (A) loosely tucked, (B) swaddled and (C) firmly tucked, and firmly tucked (Wilson et al., 1999).

1.5 Conclusions and future trends

Laboratory-based physical tests on fibers, yarns, and fabrics dominated textile research during the first part of the 20th century. Since then investigations have focused more on closer representation of end applications—three-dimensional arrangements, pretreatment of items prior to determining properties, use of manikins, and human trials conducted under controlled and field conditions. We encourage investigators to consider issues raised in this chapter, particularly those relating to understanding and accounting for variability, understanding the type of data being collected and thus appropriate design and forms of analysis, and the need to carry out sufficient pretesting to determine parameters of the investigation. These steps can enhance understanding and stimulate further development of textile products for different end uses. Investigative approaches will continue to be varied—experiments, surveys, case studies, and histories.

With textiles being so much a part of international trade, we expect increased interest in compliance with international standards, development of further international/regional agreement on methods, and development and implementation of processes leading to authentication (organic, biodegradable, environmentally friendly, and country of origin). Textile artifacts will continue to reveal rich histories.

1.6 Sources of further information and advice

1.6.1 Book chapters

Chapter 2: Sampling and statistical analysis in textile testing (Choi, 2008)
Chapter 4: Application of statistical methods in evaluation of fabric hand (Militky, 2005)

1.6.2 Test methods and websites

Many test methods have been agreed as Standard, international (International Organization for Standardization, ISO), de-facto international (agreement among a large number of countries, eg, EN), regional (Australia/New Zealand, AS/NZS), national (eg, American Society for Testing and Materials, ASTM), and product/organization specific (eg, International Wool Textile Organisation, IWTO; American Association of Textile Chemists and Colorists, AATCC).

AATCC: American Association of Textile Chemists and Colorists, http://www.aatcc.org/
AS/NZS: Australia and New Zealand Standards, http://www.standards.co.nz/; www.standards.org.au/
ASTM: American Society for Testing and Materials, http://www.astm.org/
BS: British Standards Institution, http://www.standardsuk.com/
ISO: International Organization for Standardization, http://www.iso.org/
IWTO: International Wool Textile Organisation, http://www.iwto.org/
EN: European Standards (European Committee for Standardization—CEN, Comité Européen de Normalization), http://www.cencenelec.eu/standards
Oeko-tex: Oeko.tex Association, https://www.oeko-tex.com

Acknowledgments

Part of this chapter was stimulated by teaching and supervision of students in Clothing and Textile Sciences at the University of Otago. The authors are grateful to colleagues (including

Dr. D. Carr, Dr. B. Lowe, Dr. C. Smith), and postgraduate students (including Ms. S. Gore, Ms. E. Holland, Dr. J. Malthus, Mr. B. MacRae, Dr. R. McQueen, Dr. J. Webster) whose work has contributed examples for this chapter.

References

Choi, K.F., 2008. Sampling and statistical analysis in textile testing. In: Hu, J. (Ed.), Fabric Testing. Woodhead Publishing Limited in Association with The Textile Institute, Cambridge, UK.

Cochran, W.G., Cox, G.M., 1992. Experimental Designs. John Wiley, New York.

Concise Oxford English Dictionary, 2011. Oxford University Press, Oxford UK.

Davies, O.L., 1979. Design and Analysis of Industrial Experiments. Longman, London.

European Committee For Standardization, 1999. EN 12751: Textiles—Sampling of Fibres Yarns and Fabrics. European Committee for Standardization, Brussels.

Gacula, M.C., Singh, J., 2009. Statistical Methods in Food and Consumer Research. Elsevier/ Academic Press, Amsterdam/Boston, MA.

Gore, S.E., Laing, R.M., Wilson, C.A., Carr, D.J., Niven, B.E., 2006. Standardizing a pre-treatment cleaning procedure and effects of application on apparel fabrics. Text. Res. J. 76, 455–465.

Hair, J.F., Black, W.C., Babin, B.J., Anderson, R.E., 2010. Multivariate Data Analysis: A Global Perspective, seventh ed. Prentice Hall, Upper Saddle River, NJ.

Harraway, J., 1993. Principles of experimental design. In: Introductory Statistical Methods and Analysis of Variance. second ed. University of Otago Press, Dunedin.

Hartigan, J.A., 1972. Direct clustering of a data matrix. J. Am. Stat. Assoc. 67, 123–129.

International Organization for Standardization, 2005. ISO 139: 2005 Textiles—Standard Atmospheres for Conditioning and Testing. International Organization for Standardization, Geneva, Switzerland.

Kim, J.J., Yoo, S., Kim, E., 2005. Sensorial property evaluation of scoured silk fabrics using quad analysis. Text. Res. J. 75, 418–425.

Laing, R.M., Holland, E.J., Wilson, C.A., Niven, B.E., 1999. Development of sizing systems for protective clothing for the adult male. Ergonomics 42, 1249–1257.

Laing, R.M., Tan, S.T., McDouall, J., Wright, C., Niven, B.E., Wilson, C.A., 2002. Pretibial injury in patients aged 50 years and over. N. Z. Med. J. 115, 1–11.

Laing, R.M., Niven, B., Bevin, N., Matthews, M., Wilson, C., 2006. High visiblility, UVR protection and passive cooling integrated in workplace clothing. J. Occup. Health Saf.—Aust. N. Z. 22, 567–578.

Laing, R.M., Carr, D.J., Wilson, C.A., Tan, S.T., Niven, B.E., Davis, C., Bialostocki, A., 2008. Pretibial injury: key factors and their use in developing laboratory test methods. Int. J. Low. Extrem. Wounds 7, 220–234.

Leedy, P.D., Omrod, J.E., 2016. The experimental study. In: Practical Research—Planning and Design, eleventh ed. Pearson, Boston, MA.

MacFie, H.J., Bratchell, N., Greenhoff, K., Vallis, L.V., 1989. Designs to balance the effect of order of presentation and first-order carry-over effects in Hall tests. J. Sens. Stud. 4, 129–148.

MacRae, B.A., Laing, R.M., Niven, B.E., Cotter, J.D., 2011. Pressure and coverage effects of sporting compression garments on cardiovascular and thermoregulatory function and exercise performance. Eur. J. Appl. Physiol. 112, 1783–1795.

Malthus, J.E., 1996. European Women's Dress in Nineteenth-Century New Zealand. PhD Thesis, University of Otago, New Zealand.
Manly, B.F.J., 2005. Multivariate Statistical Methods. A Primer, third ed. Chapman and Hall, London.
Maslen, G., 2013. Major research council opts for open access policy [Online]. Available: http://www.universityworldnews.com/article.php?story=2013011009450052 (accessed 25.02.13).
McQueen, R.H., Laing, R.M., Brooks, H.J.L., Niven, B.E., 2007a. Odor intensity in apparel fabrics and the link with bacterial populations. Text. Res. J. 77, 449–456.
McQueen, R.H., Laing, R.M., Wilson, C.A., Niven, B.E., Delahunty, C.M., 2007b. Odor retention on apparel fabrics: development of test methods for sensory detection. Text. Res. J. 77, 645–652.
McQueen, R.H., Laing, R.M., Delahunty, C.M., Brooks, H.J.L., Niven, B.E., 2008. Retention of axillary odour on apparel fabrics. J. Text. Inst. 99, 515–523.
Militky, J., 2005. Application of statistical methods in evaluation of fabric hand. In: Behery, H.M. (Ed.), Effect of Mechanical and Physical Properties on Fabric Hand. Woodhead Publishing Limited in Association with The Textile Institute, Cambridge, UK.
Miller, R.W., 2002. Subjective property characterization by 'Quad' analysis: an efficient method for conducting paired comparisons. Text. Res. J. 72, 1041–1051.
New Zealand History, 2015. [Online] Available: http://www.nzhistory.net.nz/politics/treaty-of-waitangi (accessed 02.04.16).
Noppa, H., Andersson, M., Bengtsson, C., Bruce, A., Isaksson, B., 1980. Longitudinal studies of anthropometric data and body composition—the population study of women in Goteburg, Sweden. Am. J. Clin. Nutr. 33, 155–162.
Pheasant, S., Haslegrave, C.M., 2005. Bodyspace: Anthropometry, Ergonomics and the Design of Work, third ed. CRC Press/Taylor and Francis, Boca Raton, FL.
Siegel, S., Castellan, N.J., 1988. Nonparametric Statistics for the Behavioral Sciences, second ed. McGraw-Hill, New York.
Sudman, S., Bradburn, N.M., Blair, E., Stocking, C., 1977. Modest expectations: the effects of interviewers' prior expectations on responses. In: Survey Design and Analysis. Sage Publications, Los Angeles, CA.
The Textile Institute, 2015. Textile terms and definitions [Online]. Manchester, UK. Available: http://www.ttandd.org/ (accessed 26.06.15).
University Of Otago, 2015a. Guidelines for ethical practices in research and teaching involving human participants. [Online] Available at http://www.otago.ac.nz/administration/policies/otago029484.html (accessed 23.02.15).
University of Otago. 2015b. Research consultation with Maori policy [Online]. University of Otago, Dunedin. Available at http://www.otago.ac.nz/administration/policies/index.html?policy=Research (accessed 23.02.15).
Webster, J., Laing, R.M., Enlow, R.L., 1998. Effects of repeated extension and recovery on selected physical properties of ISO-301 stitched seams. Part 2: Theoretical model. Text. Res. J. 68, 881–888.
Wilson, N.C., Russell, D.G., Paulin, J.M., Hopkins, W., Horwath, C., Mann, J., Nye, T., Parnell, W., Spears, G., Wilson, B., Worsley, T., Cushman, G., Laidler, A., Birkbeck, J., 1990. Life in New Zealand Survey summary report. University of Otago, Dunedin, New Zealand.
Wilson, C.A., Taylor, B., Laing, R.M., Williams, S.M., Mitchell, E.A., New Zealand Cot Death Study Group, 1994. Clothing and bedding and its relevance to sudden infant death syndrome in New Zealand infants. J. Paediatr. Child Health 30, 506–512.
Wilson, C.A., Niven, B.E., Laing, R.M., 1999. Estimating thermal resistance of the bedding assembly from thickness of materials. Int. J. Cloth. Sci. Technol. 11, 262–276.

Wilson, C.A., Laing, R.M., Niven, B.E., 2000. Multiple-layer bedding materials and the effect of air spaces on 'wet' thermal resistance of dry materials. J. Hum. Environ. Syst. 4, 23–32.
Wilson, C.A., Laing, R.M., 2002. Estimating thermal resistance of dry infant bedding: a theoretical mathematical model. Int. J. Cloth. Sci. Technol. 14(1), 25–40.
Wilson, C.A., Laing, R.M., Tamura, T., 2004. Intrinsic 'dry' thermal resistance of dry infant bedding during use: Part 2: estimated vs. measured. Int. J. Cloth. Sci. Technol. 16, 310–323.

Section B

Performance testing for specific applications

Comfort testing and fit analysis of military textiles

2

A. Schmidt, R. Paul, E. Classen, S. Morlock, J. Beringer
Hohenstein Institute for Textile Innovation gGmbH, Boennigheim, Germany

2.1 Introduction

Clothing is the first line of defense for the wearer, not only in an emergency situation, but also in their daily routine. Occupational hazards determine the extent of protection that clothing has to offer. Protective textile materials can be broadly classified as: impermeable, semipermeable, air permeable, and selectively permeable materials. Although the majority of commercially available protective textiles available for end users consist of impermeable clothing, protection offered by impermeable clothing often comes at the expense of breathability and other comfort characteristics of textiles. This eventually results in high levels of discomfort and extreme physical stress for the wearer.

The deployment of soldiers to areas with extreme climates has put new demands on the functions of military clothing for weather protection and comfort. Nowadays, military clothing plays a key role in protecting soldiers and it must perform several functions in the most rugged conditions, and has to be adapted to different fields of operation. At the same time, it is highly desirable that the clothing remain durable and above all, comfortable.

In combat situations, ballistic protective vests should protect from bullets and shell splinters. Therefore a high protection level is the main target in development. On the other hand, wear comfort of such vests is limited, especially when the wearer sweats, due to heat and/or increased physical activity. This unsatisfactory wear comfort is a risk for soldiers too, eg, when feeling too warm, or clothing is scratching the skin, they will be deflected. Furthermore, too warm clothing systems reduce physical fitness, and in extreme cases, can cause hyperthermia.

The functional properties added to military clothing are aimed at providing increased safety and survivability of soldiers in the field. Military clothing systems normally have to resist greater strains than normal garments. Materials tend to age much faster in cold conditions and under extreme exposure to the sun. Successful missions nowadays often depend on the performance of the tactical clothing and equipment. The soldier's survivability, mobility, and stamina are closely linked with wear comfort and the fit of the clothing. For wearers who perform high physical activity, it is vital that the military clothing system protects from the impact of the environment and that heat and moisture is led away from the body. A perfect balance between performance, comfort, and high durability is required. In this respect, several critical tests are necessary to ensure that

Performance Testing of Textiles. http://dx.doi.org/10.1016/B978-0-08-100570-5.00003-7

strict performance requirements are met and, ultimately, the soldier is properly protected to ensure efficiency and effectiveness of military operations.

The body reacts to temperature changes with changes in energy use leading to reduced performance of soldiers. Optimal clothing supports the soldier by providing physiological function. This means that in a specific climate/activity scenario, properly designed clothing must assist human thermoregulation in such a way that the body core temperature remains at a steady value between 36.5°C and 37.5°C. The human body generates heat energy at a steady state "metabolic rate." It varies from 80W while sleeping up to 800W or more in very high physical activity. To maintain the body core temperature constant at about 37°C within a limit of only ±2°C at varying metabolic rates, the human body has its own thermoregulatory mechanism as shown in Fig. 2.1. Additionally, controlling the humidity in the "microclimate," ie, in the air layer within the clothing next to the skin, is extremely important to maintain values low enough to be subjectively perceived as comfortable.

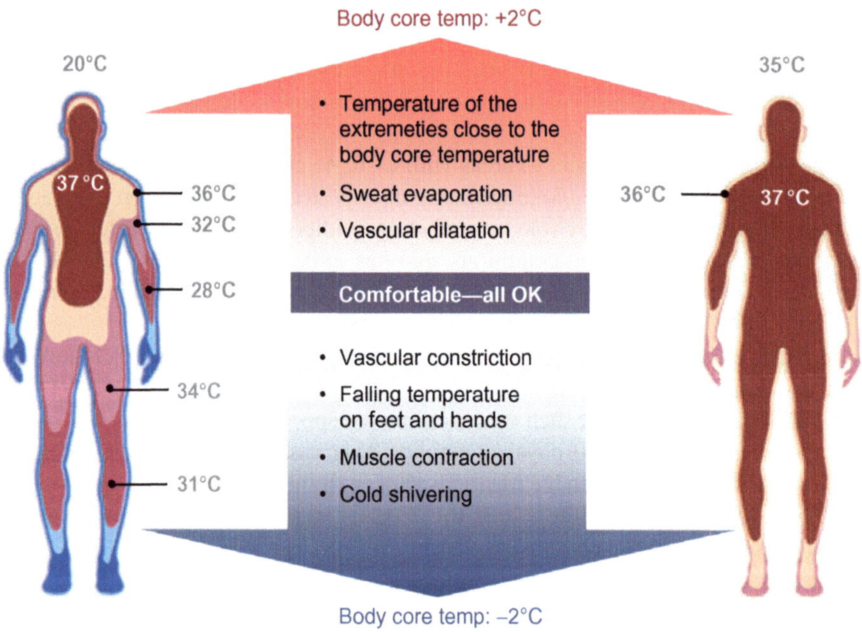

Fig. 2.1 Thermoregulatory mechanism of the human body. © Hohenstein Institute.

2.2 Comfort testing of military clothing

In general, the military specifications are normally set at higher levels or require specific modifications. Military textiles are normally tested and certified for textile ballistic resistance, textile safety, UV protection, antimicrobial effectiveness, fire resistance, and many other technical features. The wear comfort, fit, and design are equally important and this chapter focuses on these areas.

2.2.1 Thermophysiological properties

Wear comfort is essential for any successful military operation. Nowadays armed forces must be prepared to resolve conflicts anywhere in the world, from blistering deserts to frigid wastelands. Soldiers are the most important and most vulnerable part of a mission and are likely to encounter severe heat stress that must be managed for successful mission accomplishment.

Heat stress results from the interaction of climatic conditions, body heat production, and clothing or equipment that impedes heat loss. Tactical clothing often shows high thermal insulation and low permeability to water vapor, which greatly impedes heat loss, and thus exacerbates heat stress. Dedicated research on clothing physiology has resulted in developing innovative clothing systems that can reduce susceptibility to heat injury considerably.

By definition, clothing physiology is the interaction of body, climate, and clothing. With scientific methods the impact of clothing on the well-being of the wearer and wear comfort in textiles can be measured and quantified. In order to assess the physiological properties of textiles during design, among other factors, breathability has to be measured and expressed accurately and be reproducible. Such assessments must correlate to the physiological comfort perceived objectively and subjectively by the wearer.

Clothing physiology, the science and engineering of physiological function and the wear comfort of textiles and garments, became a quantitative science some 50 years ago (http://www.hohenstein.de/media/downloads/FC_EN_Bekleidungsphysiologie_mail.pdf). Over this time, scientific methods and apparatus have been developed to measure the water vapor transmission of textiles, which is critical to comfort and performance. These methods differ in the principle of the test, the complexity and duration of the test procedure, as well as the cost for the test apparatus. Among others, there are two primary methods and models that have emerged for testing the breathability of textiles. One method uses the so-called sweating guarded hot plate test (Skin Model), determining the water vapor resistance as R_{et} in $m^2 Pa/W$. This method is internationally standardized in ISO 11092 or EN 31092. The second recognized method utilizes a so-called Cup Test determining the water vapor transmission rate (WVTR) g/m^2 24 h, nationally standardized in ASTM E 96.

The Hohenstein Institute in Germany has played a crucial role in shaping the clothing physiology of textile products. The Hohenstein Skin Model (sweating guarded hot plate) is one of the tools for the measurement of the thermal and vapor resistance of textile materials, both of which are fundamental to thermophysiological comfort. In order to determine the thermophysiological quality of textile materials, the Hohenstein Skin Model (Fig. 2.2) simulates the way the skin emits heat and moisture. The measurements supply precise data on thermal insulation and moisture transport resistance, perspiration transport, sweat buffering, drying time, etc.

Using a porous sintered metal plate as a measuring surface, water vapor and fluid water are released in a controlled manner in a climatic chamber, thus simulating perspiration of human skin and different wear situations with different levels of sweat production. Moisture sensors between the measuring surface and the textile to be

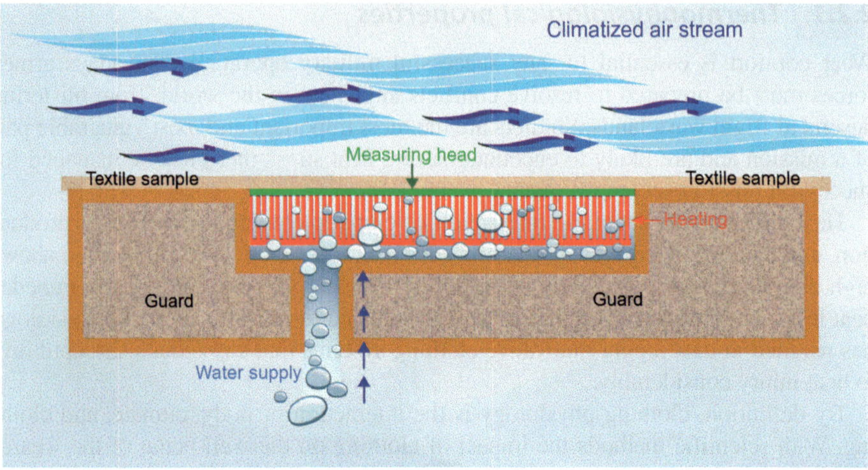

Fig. 2.2 Hohenstein Skin Model. © Hohenstein Institute.

tested measure the buffer effect of the textile and how much water vapor can be transported from the body within a specific time. This measuring technique supplies more accurate and detailed results. While the manual evaluation of the measuring data from a series of wear tests could previously take up to three months, the computer-based system can complete this task in a matter of a few hours. Measurements using the Hohenstein Skin Model now form standard tests in the field of clothing physiology worldwide. In Germany, Standard DIN 54101 has defined the measuring method using the Hohenstein Skin Model since 1991. This was replaced in 1993 by the international standards EN 31092 and ISO 11092.

Another method to test water vapor transmission of materials is the cup test according to ASTM E96—Procedure B. In this method, a cup is filled about halfway with water. The sample is placed over the top of the cup and sealed with wax on the sides to allow the water vapor to evaporate through the sample. All the samples will be run at 50% relative humidity and 23°C. The cup is placed on a rotating platform in a conditioned chamber and run for 24 h unless otherwise specified. The samples are weighed periodically during the 24h period and a best fit line is achieved. A water/moisture vapor transmission rate is calculated in g/m^2/24 h (Huang and Qian, 2008).

2.2.1.1 *Skin Model (sweating guarded hot plate) versus cup test*

In comparing the two methods for testing the breathability of textiles, from scientific, technical, and practical points of view, the Skin Model has significant advantages over the cup test. The first advantage is that the Skin Model allows for precisely defined and controlled test conditions, thus yielding more accurate and reproducible results. However, the most important difference between the methods may be that various research projects have demonstrated that the R_{et} value measured with the Skin Model correlates directly with the wearer's skin wetness or the garments' upper limit of the temperature range of utility (TRU). TRU for a specific activity level is the maximum temperature where the wearer is not feeling uncomfortable due to moisture effects. When the WVTR value is measured with the cup method however, no such correlation to wearer comfort can be determined. Therefore, using the Skin Model, the actual interaction between the body and textile garments is far better simulated than with the cup test.

2.2.2 Thermal insulation

The thermal insulation of military garments, combat uniforms, made-up goods, immersion suits, cold protective clothing, duvets, and sleeping bags can be measured with the help of a thermally segmented testing manikin. "Charlie," the thermal manikin developed by the Hohenstein Institute is shown in Fig. 2.3. The thermal manikin consists of 16 segments, which can be heated separately. Using what are known as human thermoregulation models, the heat generated by adults is set. The segmented manikins are made of copper or synthetic materials and have been fitted with a computer-controlled heating system that allows the heat generation for different parts of the body to be regulated individually and independently of one another. The more heat emitted from the arms or legs, for example, the worse the thermal insulation of the garment is for those areas of the body. These figures are very significantly influenced by the movement of air when the body is in motion. Therefore the latest version of the segmented manikin, "Charlie 4" has been set up so it is able to move during testing, as if it is out for a brisk walk.

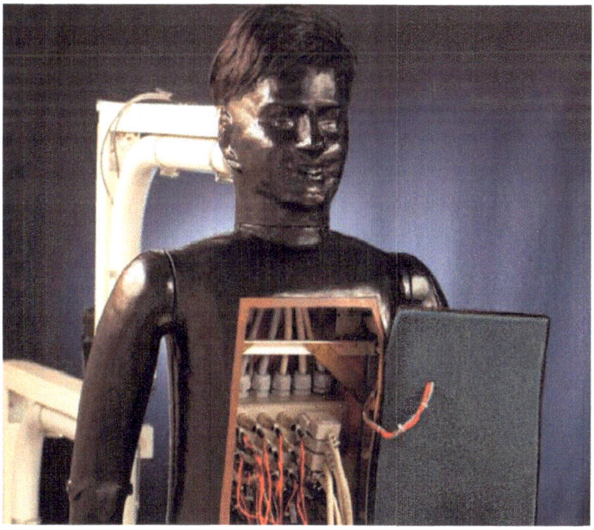

Fig. 2.3 Thermally segmented testing manikin. © Hohenstein Institute.

The assessments made using the thermally segmented testing manikins are an important complement to those made using the Skin Model, because the influence of the way the item or garment is made (fit, elasticized cuffs, turtlenecks, etc.) can be taken into consideration. The measurements are carried out using the thermal manikin according to EN ISO 15831. During testing, it is dressed with a complete clothing system. To measure the thermal insulation of this system, the manikin is placed in a climatic chamber either in a standing position and/or is set to move. To provide for realistic testing the most varied of environments can be created in the climatic chambers. A precipitation system can be used to create rain of a wide range of intensities, and an IR radiant wall simulates intense sunshine or heat. A wind generator can deliver wind speeds up to 10 m/s.

Wear trials can be performed to investigate the influence of the measured values on the wear comfort of a real person. Those wear trials take place in a climatic chamber under standardized conditions. Heart rate, body core temperature and skin temperature were recorded as well as temperature and relative humidity in the microclimate between skin and clothing. Furthermore, subjects have to state their perception of temperature, humidity, and wear.

2.2.3 Skin sensorial comfort

In addition to thermophysiological wear comfort (heat and moisture management), the skin sensory wear comfort is a key determinant of comprehensive wear comfort for a textile that is worn in contact with the skin. The skin sensorial properties are quantified where the number of contact points between textile and skin shows whether a textile feels clammy or sticky on the skin. The skin sensorial tests are performed according to in-house standard operation procedures with standardized and accredited test methods. These test procedures have been developed by the Hohenstein Institute based on numerous wearer trials with test persons and decades of experience with testing skin sensorial properties of textiles. According to the intended use of the textile, the values and their weight is adjusted and the formulas are different for military clothing, sports clothing, and casual wear. The test results are expressed as skin sensorial comfort vote, and in order to calculate it, the textiles have to go through five tests which provide information about both their skin sensorial and their thermophysiological characteristics.

2.2.3.1 Stiffness

The stiffness of a textile is an indicator of how well a textile will adapt to the body shape. It is measured in a special device by using a laser beam to calculate the bending angle of a sample strip draped over a thin stick. Based on in-house expertise, the Hohenstein Institute has defined the criteria for different products and areas of use which ensure maximum wearing comfort and prevent any mechanical skin irritation caused by excessive stiffness.

2.2.3.2 Sorption index

The sorption index uses time and contact angles to measure how quickly a water droplet (ie, sweat) is absorbed by the textile. As the skin becomes more moist, it becomes more sensitive to mechanical irritation. That is why it is important for a textile material to transport sweat away from the skin as quickly as possible. The sorption index indicates the speed at which a drop of water on the textile is absorbed by it. To measure this, a drop of water is applied to the textile sample and observed by a video camera. The angle of contact between the drop of water and the surface of the textile is measured continuously to establish how quickly the material absorbs liquid sweat.

2.2.3.3 Surface index

The surface index expresses the hairiness or roughness/smoothness of a textile. Here, a camera records an image of a cross-section of the textile. The number and size of the protruding fibers in the fabric can then be calculated. The surface index can be used to judge, for example, whether a textile will scratch or feel too smooth.

2.2.3.4 Number of contact points

The number of contact points between the textile and skin states how fast a textile will be sensed as clammy or damp. Image analysis systems connected to a surface scanner show the number of contact points and the surface structure of textiles. These can be used as a scale for working out what area of the textile material is in contact with the skin. Based on in-house expertise, the Hohenstein Institute has developed guidelines for the optimum number of contact points for textiles.

2.2.3.5 Wet cling index

The wet cling index indicates how likely a textile is to adhere to perspiration moistened skin. Water is applied to a sintered glass plate to simulate sweating skin. The textile sample is mounted on a specimen holder and drawn over the plate. The force required to do this is measured and it serves as the basis for the wet cling index, from which it can be judged whether the textile will stick to the skin and feel uncomfortable when one sweats.

In general, the skin sensorial comfort is negatively affected by hydrophobic and smooth surfaces that easily cling to sweaty skin, or which tend to make textiles stiffer. The use of spun yarns and hydrophilic treatments improves the skin sensorial wear comfort. These tests are especially important for military clothing treated with insect repellent or flame retardant finishes, as these finishes, while providing functional properties to textiles, may negatively affect their feel and handle.

2.2.4 Evaporative cooling power of garments

In hot operation areas one of the most important issues is to prevent the soldier from hyperthermia and ultimately death by heat stroke. Therefore the excess heat generated by movement and carrying heavy combat gear has to be dissipated through the whole garment system. The "air conditioner" of the human body dissipates excess heat by the evaporation of liquid sweat—the so-called evaporative heat loss. To increase that evaporative heat loss, special cooling garments can be worn close to the skin as functional underwear. These cooling garments can be either active or passive.

- Active cooling garments pump cold water through a tube system integrated in, ie, a vest dissipating the heat away from the body, similar to a water cooling system in engines. For this technology an additional technical device with pumps and a control unit is needed, which is usually carried as a backpack and generates additional weight.

- Passive cooling garments do not need these additional technical devices. They simply generate a cooling effect by enhanced sweat evaporation through a smart combination of new polymers as well as innovative yarn and fabric construction.

Various cooling garments with said cooling effect under distinct environmental conditions are on the market or also under development. Their cooling properties are broadly used for the marketing of these fabrics and garments. However, this is usually done without objectively measuring data about the achieved cooling power and resulting temperature drop between the skin and the textile.

The cooling effect data is often based on testimonials and trials in which subjects wear the garments and state afterwards that there was a cool feeling on the skin. This is of course also due to the fact that there was no characterization method available which is able to technically prove the cooling effect like the wearer really feels it when wearing the textile. This can now be done with the Hohenstein measuring system, WATson.

The basic principle of WATson is a heated plate (measuring head) which is held constant at skin temperature by electronically controlled electrical heating, hence simulating the heat release of the human skin and body. By measuring the needed electrical power to keep the measuring head at set temperature when a "cold" fabric comes into contact, the cooling power can be directly determined. Result is the cooling power in watts per area over time. By placing temperature sensors on and under the fabric, the perceived temperature over time of the garment on the skin can also be determined. Furthermore, the WATson setup can be exposed to various ambient climates (cold/dry to hot/humid) as well as numerous edge conditions (sweating, wind, and IR radiation) to quantify the thermal properties of single cooling garments and also whole garment systems in real life conditions.

WATson is based on the principles of the Hohenstein Skin Model but is an all new and separate development. It can quantify the cooling effect of garments at a very small sample size of only 25 × 25 cm and so no full garment is needed for this test. By this it can be now technically proven with measured data if a cooling effect really takes place, how much cooling power is generated, how long this effect lasts and therefore how long the soldier can be kept cool during a mission in hot operation areas.

2.3 Comfort testing of other textile products

2.3.1 Extreme cold protective clothing

For cold protective clothing, a low water vapor resistance is of high importance because it leads to reduced sweat accumulation in the clothing system. This is not only the case for "normal" cold temperatures of 20°C but also at freezing point and far below at −20°C. Excess energy has to be dissipated by sweat evaporation, and an energy loss in cold environments has to be compensated by shivering. Both excessive sweating and shivering result in losses in human performance efficiency, so it is desirable to control these consequences through better protective clothing with appropriate thermal insulation. The challenge for military clothing designers is to achieve an even energy balance in the clothing system by understanding and balancing breathability and insulation.

The thermal insulation of military uniforms in cold environments depends on the ambient temperature and the metabolic rate of the soldier. A good example for this is the European standard EN 342, which measures thermal insulation for cold protective clothing. In this standard, the thermal insulation of the garment is measured for a maximum wearing time at a certain metabolic rate under known ambient temperature. This information generates the criteria required for optimal performance. Thermal insulation (ie, thermal resistance) may also be determined on a guarded sweating hotplate like Skin Model according to EN 31092/ ISO 11092. This can be used for fabric measurement and design. When testing whole garments and/or whole clothing systems, thermal manikins according to ISO/DIN 15831 are used.

Thermal insulation itself mainly depends on the enclosed air volume in the garment, as air has a very low thermal conductivity and is a good insulator. However, when a garment gets saturated and wet, the insulating air is replaced by humidity first as water vapor and then as water, which in contrast is a very good conductor of heat and a much poorer insulator. Therefore, body heat is lost rapidly under wet conditions. To keep a garment dry on the inside while sweat is evaporated by the wearer due to increased physical activity, a low water vapor resistance is essential. To evaluate a fabric in regard to the water vapor resistance, the Skin Model according to EN 31092/ISO 11092 is also used. The lower the water vapor resistance is, the higher the breathability is.

As a final test sequence, to evaluate all performance aspects of a military uniform for a cold environment, subject wearing trials are conducted in a climatic chamber that simulates realistic temperatures and physical activity/metabolic rates. The subject is equipped with numerous temperature and humidity sensors on the skin that judge his or her subjective thermal and moisture sensation as well as the resulting overall comfort in distinct time periods and varying conditions. The subjective perceptions of the wearer and the measured temperature and moisture data are compared and subsequently correlated with the data from the Skin Model and thermal manikin. This fully validates the wear trials.

Military uniforms for cold environments can be designed for specific climatic conditions and physical activity using existing test procedures and standards. The thermal insulation and the water vapor resistance of the whole garment are important. A low water vapor resistance ensures sweat evaporation and results in a dry insulation layer, and thus less heat loss. The improved overall comfort ensures the mental and physical performance of the soldier in extremely cold environments.

2.3.2 Duvets and sleeping bags

Clothing physiology is not only important for military uniforms, but also for bed linens and duvets for improving the sleep comfort of soldiers. Apart from such "comfort" reasons, a shivering soldier is not able to successfully aim at a target and is endangered by hypothermia as well. So adequate rest and sleep are necessary for soldiers who are in the outdoors for long periods of time. For sleeping in the outdoors, thermal mats and sleeping bags are generally used. Particularly important is the sleeping bag, because of its direct contact with the sleeper's body. Therefore, military forces should not fail to provide optimized sleeping bags for their troops, because physical recovery of the soldiers may turn out to be a crucial factor in field operations.

The thermal insulation of bedding, duvets, and sleeping bags can be measured with the help of a thermally segmented testing manikin. When evaluating the sleep comfort of duvets, the focus is on the temperature balance of the human body in relation to the environmental temperature and the heat and moisture management within the "sleeping cavern." A physiological sleep comfort rating ranging from 1 (= excellent) to 4 (= unsatisfactory) can be calculated for duvets. Similar to the wear comfort rating for clothing, this makes a statement on the ability of the product to maintain a pleasant body temperature when in bed or asleep, and to quickly and effectively transport any sweat away from the body. The thermal insulating effect of a duvet plays a particularly important role in temperature management.

Sleeping bags should optimally support the thermoregulation of the soldier in any environment. A sleeping soldier's body produces heat—the equivalent of an 80W heater. However, this heat is lost via convection, conduction, sweat evaporation, radiation, and respiration. In a comfortable sleeping bag, heat production, and heat loss are in equilibrium. This produces optimum sleeping comfort. If heat generation and heat loss are not in equilibrium, soldiers may sleep poorly, or in the worst case, in colder environments, frost bite and hypothermia may occur.

Since 2002, the thermal insulation of civil sleeping bags, which are produced for the European market, has to be measured according to EN 13537 using a thermal manikin. This testing uses a sophisticated physiological model allowing prediction of temperature differential performance, from actual thermal insulation measurements of the sleeping bag materials. These data define a certain "zone" of utility for the sleeping bag: including a comfort, transitional, and risk zone.

The standard testing method EN 13537 states that it does not apply to sleeping bags intended for military use, because of the differences of physical condition and physiology between soldiers and civilians. However, thermal insulation is measured with high accuracy and sleeping bags can be clearly distinguished by it. Therefore, such testing does indicate which sleeping bags can be chosen in regard to the anticipated use environment.

To get a more realistic, military use scenario, testing must move beyond the conditions of EN 13537. A test manikin should be bedded on a military thermal mat on the ground or even on earth, and since soldiers do not necessarily sleep in a tent, higher air flows could be simulated by means of wind generators to provide a more realistic testing condition. In addition, the physiological model used for the calculation of the temperature limits has to be adapted, because soldiers have a higher heat production, allowing for higher limits for hypothermia.

2.3.3 Immersion suits

Apart from the danger of drowning, death from hypothermia is another great danger for offshore military missions. It only takes a few minutes in cold water for a soldier to lose his ability to move about without restriction. It is crucial that the soldiers are equipped correctly when it comes to the thermal protection of immersion suits. Hohenstein is one of the few institutes worldwide that are in a position to test the thermal insulation and survival time in immersion suits with a thermal manikin according to ISO 15027 without the need for expensive and ethically not justifiable real life testing. ISO 15027-1:2012 specifies performance and safety requirements for constant wear immersion suits for work and leisure activities to protect the body of a user against the effects of cold water immersion, such as cold shock and hypothermia. It is applicable for dry and wet constant wear immersion suits. The measurements are carried out in calm but moving water. The climatic chamber (−20°C to +40°C) and a wind generator (with wind speeds up to 10 m/s) enable customized test designs as a wide range of environmental conditions can be simulated.

Helicopter pilot suits for offshore applications often show reduced thermal comfort as well as lack of ventilation. This thermal stress, also induced with the high cockpit temperatures, may lead to reduced time to fatigue and thus occurrence of accidents. The Hohenstein thermal manikin "Charlie" and the Skin Model allow for an objective assessment and optimization of breathability and thermal resistance of the same.

2.3.4 Military vehicle seats

Troops should not only arrive safely, but also arrive in the best condition. Seats often still contain layers, which are easy to care for, but are not able to absorb and transport humidity and heat away from the surface layer. This causes heat and moisture accumulation resulting in an uncomfortable seat climate. It is scientifically proven that these issues lead to the increased fatigue of drivers and thus risk of accidents. To test and further develop different seating materials, the newly developed Hohenstein Sweating Bottom allows dynamic measurements as it simulates driving movements and provides statements on moisture transport, water vapor resistance, and buffering capacity of water vapor/vaporous sweat. The state-of-the-art model enables dynamic measurements as it simulates the movements of the seat during actual driving. The special design, including the bottom and part of the thigh, allows conclusions regarding the optimal make-up of the upholstery.

2.3.5 Combat boots and gloves

Optimized shoe-sock systems can prevent feet from overheating, and bacterial and fungal infections, and thus help to increase soldiers' performance. The thermal insulation and breathability of combat boots, gloves, and socks can be evaluated with the thermal regulation model, the "Sweating Foot." To achieve fit and function, the Sweating Foot enables a realistic simulation of the special thermal characteristics of human extremities, and delivers reliable and differentiated results on thermal insulation and breathability.

2.4 Fit of military clothing

In civil life, fit of clothing is mostly a question of good looks. For soldiers, ill-fitting and uncomfortable clothing can be irritating at best, but deadly at worst. Crawling soldiers in uniforms that are too big are at significantly higher risk of getting caught. On the other hand, too tight clothing will decrease their ease of movement.

Furthermore, the proportions of the average human body have considerably changed during the last decades. People are growing taller and have increased in weight. Therefore it is necessary to use representative size surveys to redefine clothing sizes. A special challenge, which public services such as military, police, and fire brigades are facing worldwide, is the uprising number of women on duty. So far, uniforms and protective clothing have mostly been designed to fit men. Women wearing these badly fitting uniforms, however, are less protected, as current studies on the protection level of female fire fighters have shown.

In this context, in 2007, Hohenstein Institute and the Human Solutions GmbH have started a joint project named SizeGERMANY. The body measurements gathered during the project provide the basis for what is known as the Hohenstein Sizing Chart. The obtained data also served as a calculation basis for the new "standard man and woman." Combined with collected Caucasian sociodemographic data, it is possible to define certain target groups, for example female soldiers, and the percentage of any size for this group.

The knowledge on the "standard soldier" and the required number of pieces for the individual size groups on stock, can help to reduce procurement costs in the military sector. More detailed analysis would also allow reducing the number of special combat uniforms or personal protective equipment, which is not necessary for every soldier. Another field which could benefit from the gained data is the ergonomic design of seating. In order to minimize the stress of getting to the combat zone, seating of drivers and pilots could be improved by means of scans in different postures.

2.4.1 Fit and head protection

One of the most critical parts of the personal protective equipment of soldiers is still the helmet. Statistics show that in today's military operations, traumatic brain injury is the number one cause of death or disability of soldiers. Surveys among soldiers in Afghanistan revealed that most helmets were ill-fitting and uncomfortable. As a result, combat troops on patrol or convoy took their helmets off or released the chins strap for pain relief and were thus exposed to increased risk of injury.

Not every helmet is right for everyone, as all heads are different. It is a known fact that shape and circumference can differ greatly from person to person. Helmet manufacturers already know this, but they have to work with the measurement information currently available for helmets. Even the safest helmet cannot offer optimum fit and protection if it is not right for the individual head size and shape of the wearer. Despite the great demand for suitable head protection systems, there are no well founded anthropometric head data available to date. Hohenstein Institute has carried out research on this topic and the results can enable manufacturers to offer helmets with a better fit in the future. Representative pools of raw scan data from male and female test subjects were collected as a basis for defining characteristic 3D head morphologies, creating average head shapes and clearly deriving all relevant parameters for the development of corresponding protection systems. The results can be transferred to all head protection systems ranging from police and military, to construction workers, to helmets for sports enthusiasts.

In addition to the known head measurements from the ergonomics standard, additional measurements are taken and evaluated to derive characteristic head morphologies. The research project has also worked on facial measurements and provided dimensional descriptions, meaning possible dimensional correlations between the eye, nose, mouth, and ear areas. The categorizing of these morphologies into representative head shapes is an important step. Interested manufacturers can have access to specific sizing systems describing the current head and face dimensions to be used as a basis for the design of head protection systems. Standardised, realistic virtual average 3D head shapes can also be made available for commercialization.

In addition to fit-specific criteria, physiological aspects of moisture wicking have a substantial influence on the helmet's comfort. The materials used influence the climatic conditions inside a helmet. So it is important also to optimize the textile interiors of helmets with regard to clothing physiology and hygiene.

2.5 Conclusion

Textiles for military uniforms face a complex set of challenges. They must provide protection, durability, and comfort in a wide range of hostile environments. For the soldier's clothing systems, the aim is to develop lighter, more breathable and comfortable materials, with no compromise on ergonomics and operational capabilities. The military requires high levels of physical activity in a wide variety of environments, therefore military clothing systems must protect the wearer from the impacts of the operational environment and also dissipate heat and moisture. Constant improvement of the soldier's equipment is vital in order to maintain competitive advantages in rapidly changing environments and battlefield scenarios. Apart from the military clothing, all other military textile products should also provide the necessary comfort to soldiers. In addition, the fit of the military clothing and the head protection systems like helmets are also really important for successful military missions.

Reference

Huang, J., Qian, X., 2008. Comparison of test methods for measuring water vapor permeability of fabrics. Text. Res. J. 78 (4), 342–352.

Testing and evaluating the thermal comfort of clothing ensembles

3

G. Song*, S. Mandal†
*Iowa State University, Ames, IA, United States
†University of Alberta, Edmonton, AB, Canada

3.1 Introduction

Clothing is an integral part of life for all human beings and typically different types of clothing are worn every day. Clothing can be broadly classified as regular and specialized. Regular clothing is mostly suitable for temperate, ambient environments, and this type of clothing can be categorized into formal- and casual-wear. Specialized clothing is mainly required to provide protection from hazardous ambient environments (eg, high heat, extreme cold), and this type of clothing can be categorized into thermal protective clothing, chemical protective clothing, etc. Generally, human beings wear clothing to satisfy their psychological needs; however, clothing can also fulfill their physiological needs in particular ambient environments (Gilligan, 2010; Li, 2001). Although physiological needs cover a wide range of expectations (eg, less itchiness, perfect fit, thermal comfort), it is primarily required that clothing should provide proper thermal comfort to wearers under different ambient environments (Fourt and Hollies, 1970).

In order to understand the role of clothing for maintaining thermal comfort of wearers, at first, it is important to acknowledge the fundamentals of human physiology (Poehlman, 1992; Poehlman and Horton, 1989; Saris et al., 1989). Contextually, it is natural that humans eat various macronutrients—carbohydrates, proteins, and fats—to keep themselves alive; they also inhale oxygen (O_2) during respiration and drink water to keep their body hydrated. These macronutrients, water, and inhaled oxygen are carried through blood to their muscles. In the mitochondria of muscle fibers, the oxygen oxidizes macronutrients with the help of enzymes and primarily produces carbon dioxide (CO_2), water (H_2O), heat, and adenosine triphosphate (ATP). This ATP is a source of muscular energy under both resting and physically active conditions (Copper et al., 2011; Manore et al., 1991). During the process of ATP production through cell respiration, about 80% of the stored potential energy in macronutrients is converted into heat energy. If this metabolic heat cannot be dissipated through the clothing towards the ambient environment, it will tend to increase the core temperature of the human body from the standard value of 37°C (Katch et al., 2006). If an increase of 0.1–0.5°C is sensed by the central and/or peripheral nervous receptors of the body, sweat glands in the skin are activated. Sweat consists mostly of water but also contains small quantities of salts, lactate, and urea; only the water is significant from the point of view of comfort. Sweat accumulated on the human skin can be evaporated to

Performance Testing of Textiles. http://dx.doi.org/10.1016/B978-0-08-100570-5.00004-9

the ambient environment in order to keep the body temperature at 37°C. It has been observed that metabolic heat and sweat vapor generally dissipate through the clothing by the processes of radiation, convection, conduction, and/or evaporation. In some circumstances, these processes of metabolic-heat and sweat-vapor dissipation could be hindered by clothing, especially when specialized clothing (eg, thermal protective clothing, cold weather protective clothing, sportswear) is worn by workers such as firefighters, mountaineers, and sportsmen in extreme ambient environments. In these situations, a great deal of thermal discomfort may be experienced by wearers.

Considering the aforementioned role of clothing on the thermal comfort of human beings, many researchers have examined the factors that affect thermal comfort performance of clothing (Hosseini Ravandi and Valizadeh, 2011; Karaca et al., 2012; Varshney et al., 2011; Yoo and Barker, 2005). Generally, manufacturing of clothing involves a sequence of steps such as fiber processing, yarn formation, fabric formation, and/or garment manufacturing (Glock and Kunz, 2005; Mandal and Abraham, 2010). It may be expected that thermal comfort performance of clothing is closely associated with its fibers, yarns, fabrics, and methods of construction.

It has been found that thermal comfort performance of clothing primarily depends on the fabrics used to manufacture the clothing. As a consequence, thermal comfort performance of samples of fabrics has been evaluated using tests such as the sweating guarded hot plate test, which has been designed to stimulate the thermal behavior of fabrics in contact with human skin (McCullough et al., 2004; Psikuta et al., 2013). Additionally, thermal comfort performance of clothing can be directly evaluated using two methods—sweating thermal manikin tests and human trial tests (Dozen et al., 1992; Huck and McCullough, 1988; Mountain et al., 1994). In the case of manikin tests, a full-scale clothed manikin that can simulate sweating (eg, Finnish manikin “Coppelius”, Japanese manikin “Taro”, Hong Kong manikin “Walter”, United States manikin “Newton”) is used to evaluate the thermal comfort performance of clothing (Fan and Chen, 2002; Meinander, 1999; Richards and Mattle, 2001; Wu and Fan, 2009). By controlling the ambient environment conditions and posture of a sweating manikin, thermal and evaporative resistances of the clothing are evaluated under different ambient conditions. The thermal and evaporative resistance values of clothing together are used to understand the thermal comfort performance of clothing. Although standardized manikin tests from various organizations (ASTM, American Society for Testing and Materials and ISO, International Organization for Standardization) are widely used to evaluate comfort performance, it is difficult to simulate accurately the physiological aspects (eg, sweat generation, metabolic-heat production) of the human body (ASTM F 1291, 2010; ASTM F 2370, 2010; ISO 9920, 2007). Therefore, human trial methods have gained popularity in recent years to understand the thermal comfort performance of clothing. In fact, a standardized human trial method has also been developed by the ASTM (ASTM F 2668, 2011). In this human trial method, a group of human volunteers wear clothing and do physical exercise while physiological parameters such as heart rate response, maximum oxygen consumption, and body temperature are measured, to understand the thermal comfort provided by the clothing.

By using the manikin tests and human trials, many researchers have assessed thermal comfort performance of different types of regular and specialized clothing (Barker, 2008; Bhattacharjee and Kothari, 2009; Farnworth, 1983; Hes et al., 1996; Huck and

McCullough, 1988; Matusiak, 2006; Wu et al., 2009). Through their assessments, they have found that thermal comfort performance of clothing can be optimized by implementing different scientific approaches, such as selecting appropriate fabrics for clothing, designing the right fit of clothing for wearers, and controlling the ambient environmental conditions. Also, researchers have identified a need to develop state-of-the art testing standards to accurately evaluate thermal comfort performance of clothing. Furthermore, they affirm that there is a requirement to develop new clothing for providing optimum comfort to wearers.

In this chapter, different factors that affect thermal comfort performance of clothing are examined. Subsequently, various methods to evaluate thermal comfort performance of clothing will be discussed. Thermal comfort performance of different types of clothing will also be critically assessed to suggest different approaches that can improve the performance of clothing. Finally, various key issues related to thermal comfort of clothing will be highlighted to indicate future research directions. These issues are mainly directed towards development of state-of-the-art testing methods to measure thermal comfort performance of clothing and fabrication of new fabric materials for clothing that can provide better thermal comfort to wearers. Overall, this chapter will contain information to help textile and materials engineers to design clothing that can provide optimal thermal comfort to wearers under a range of different conditions.

3.2 Factors affecting thermal comfort performance of clothing

Generally, fiber-based hygroscopic or nonhygroscopic porous fabrics (woven, knitted, and/or nonwoven fabrics) are tailored into clothing to cover the complex geometry of wearers' bodies. Thus, many factors related to textile engineering (eg, properties of fibers, yarns, fabrics, garments) could affect the thermal comfort performance of clothing. Actually, all these factors affect thermal insulation characteristics of fabric and clothing, which help to achieve proper thermal comfort performance. If thermal insulation characteristics and air and water vapor permeability of clothing are low and high, respectively, the metabolic-heat and sweat-vapor from wearers' bodies will be properly dissipated, resulting in a comfortable situation for wearers. Notably, high thermal insulation and low air and water vapor permeability are essential for some types of specialized clothing (eg, thermal protective clothing for firefighters or industrial workers, cold weather protective clothing) in order to provide protection to wearers in hazardous environments (Mandal et al., 2013; Mandal and Song, 2015). The effect of fiber, yarn, fabric, and/or garment properties on thermal insulation characteristics of clothing is discussed in the following section.

3.2.1 Effect of fiber properties

Fiber is the basic element to manufacture a fabric. Fiber properties have a significant effect on thermal insulation characteristics of textiles. Important fiber properties are fineness or linear density, length, cross-section, and crimp (Matsudaira et al., 1993; Murakami et al., 1978; Varshney et al., 2011).

Fineness or linear density. Fiber fineness affects thermal insulation and is expressed in terms of denier (weight of 9000m of fiber in grams) (Li, 2001; Varshney et al., 2011). For a given weight of fiber, a smaller ratio of fiber surface to volume is available in the case of coarse fibers compared with fine fibers. As the fiber surface traps air, a given weight of coarse fiber will trap less air than a similar amount of fine fiber. Consequently, coarse fibers will generally provide less thermal insulation than fine fibers (Wu et al., 2008).

Length. It is natural that a short-length fiber will cover less surface area than a long-length fiber. This means that a short-length fiber would trap a lesser amount of air on its surface than a long fiber. Hence, a shorter fiber possesses lesser thermal insulation than a longer fiber (Marom and Weinberg, 1975; Rao and Gupta, 1992).

Cross-section. The cross-section of a fiber can be observed using a microscope. It has been found that the cross-sectional shape of a fiber can have a significant effect on its thermal insulation characteristics (Varshney et al., 2011). A fiber's cross-section with more trapped air may provide higher thermal insulation than a perfectly cylindrical fiber. For example, a hollow fiber traps more air inside its structure than a solid circular fiber. This is the reason why hollow-fiber based fabrics can provide higher thermal insulation than solid circular-fiber based fabrics. In the same fashion, a non-circular fiber, say with a trilobal or scalloped oval surface, can trap more air than a circular fiber, because of its shape. Relatively large amounts of air trapped by noncircular fibers ultimately enhance thermal insulation characteristics (Matsudaira et al., 1993; Murakami et al., 1978).

Crimp. Crimp is another important property that affects thermal insulation (Bozkurt et al., 2007; Rwei et al., 2005). It has been found that a fiber with highly crimped structure comprises a lot of loops in its surface and these loops contribute to trap air on the fiber surface. Hence, the thermal insulation provided by crimped fibers is typically higher than an uncrimped or low-crimped fiber.

3.2.2 Effect of yarn properties

When fibers are converted into yarn to manufacture woven fabrics, yarn properties (type, twist, and configuration) will also impact the thermal insulation characteristics of the resulting fabrics (Bogaty et al., 1957; Ghassemi et al., 2011; Hatch, 1993; Ramachandran et al., 2010).

Type. When comparing texturized yarns, spun yarns, rotor-spun yarns, and filament yarns, it has been found that texturized, spun, and rotor-spun yarns have protruding fibers on their surfaces. These protruding fibers trap air and increase the thermal insulation characteristics of yarns. In comparison, the surface of filament yarns is smooth; consequently, filament yarn does not trap much air on its surface. Thus, the thermal insulation characteristics of filament yarns are lower than for texturized/spun/rotor-spun yarns (Ghassemi et al., 2011; Ramachandran et al., 2010; Rengasamy and Kawabata, 2002).

Twist. It has been observed that the structure of a low-twisted yarn is less compact than with a high-twisted yarn. Due to this less compact structure, a low-twisted yarn traps more air on its surface than a high-twisted yarn. This situation results in less thermal insulation for a highly twisted yarn than a low-twisted yarn (Rengasamy and Kawabata, 2002).

Configuration. When fibers are twisted to produce a yarn, the arrangement of fibers in the yarn affects its thermal insulation characteristics (Bogaty et al., 1957; Hatch, 1993). If fibers lie nonparallel to each other when they are twisted to produce a yarn, it has been found that this type of configuration results in less thermal insulation than a twisted yarn with parallel fibers (Song, 2009). Having less air trapped inside the yarn reduces its thermal insulation characteristics.

3.2.3 Effect of fabric properties

There are various properties of fabric that affect its thermal insulation. These properties are associated with the surface, structure, design, porosity, weight, thickness, density, thermal conductivity, heat capacity, and moisture accumulation properties of fabrics (Greenwood et al., 1970; Perkins, 1979; Shalev and Barker, 1984; Song et al., 2011; Sun et al., 2000).

Surface. The surface of a fabric is one of the most important properties that affect thermal insulation. Generally, the optical property of a fiber surface (eg, color) controls its emissivity. If the surface color of a fabric is black, its emissivity is at a maximum ($\varepsilon = 1$). Alternatively, if the surface color of a fabric is white, its emissivity is at a minimum ($\varepsilon = 0$). A fabric with high emissivity absorbs thermal energy from the ambient environment and re-emits this energy, some of which travels toward wearers' bodies (Mandal and Song, 2011; Mandal et al., 2013; Shalev and Barker, 1984).

Additionally, surface roughness is another property to consider for effective thermal insulation. According to the boundary air layer theory, when moving air comes in contact with a solid surface, air may be trapped on the surface. If the surface roughness is low, it will trap less air and will provide reduced thermal insulation (Ding et al., 2011a,b).

Structure. Generally, fabric structures can be classified into three categories: woven, knitted, and nonwoven. Among these types of fabrics, nonwoven fabrics can incorporate maximum amounts of air in their structures and consequently possess the best thermal insulation properties. While nonwoven fabrics are not preferred for regular clothing, they are often used to produce thermal liners that are used in specialized clothing. A knitted fabric typically traps more air inside its structure than a woven fabric, because a knitted fabric comprises loops in its structure. In comparison, a woven fabric structure with interlacing warp and weft yarns has no loops in its structure and it traps less air than a knitted fabric of the same weight per unit area. Thus, the thermal insulation properties of woven fabrics are somewhat poorer than knitted fabrics (Celcar et al., 2008; Frydrych et al., 2002).

Design. Thermal insulation characteristics may depend upon the design of woven and knitted fabrics (Greenwood et al., 1970; Li, 2001). For example, frequent interlacement between warp and weft occurs in plain weave fabrics in comparison with, for example, a 5-harness satin weave which has many float-yarns. Due to the greater interlacement, a plain weave fabric traps more air than a satin weave fabric. This causes the thermal insulation of plain weave fabrics to be better than that of satin weave fabrics.

Porosity. Fabric porosity also affects the thermal insulation characteristics of fabrics (Fan et al., 2000). Generally, a fabric with a highly porous structure can trap more air than one with a lower or nonporous structure. The air trapped inside the fabric increases its thermal insulation.

Weight. Fabric weight is expressed in terms of gram per square meter of fabric (g/m^2). It has been found that a fabric with less weight generally can trap less air than a high-weight fabric, especially if their fiber diameters, weave structures, and thicknesses are the same (Perkins, 1979; Sun et al., 2000; Song et al., 2011). This is because, if the weave structures and thickness are the same, fabric weight can be decreased by decreasing the number of threads per unit length (warp/in. and weft/in.) in the yarns so there is less trapped air in the fabric and hence lower thermal insulation.

Thickness. Thickness is the single most important property which affects thermal insulation characteristics of fabrics because a thicker fabric always traps more air than a thinner one (Behnke, 1984; Shalev and Barker, 1983; Lee and Barker, 1986). Additionally, a thicker fabric can store more absorbed thermal energy in its structure. In general, a multilayered thick fabric can provide higher thermal insulation than a nonlayered fabric with the same thickness. This is because a fabric with a layered structure can trap more air and also provide greater space to store absorbed thermal energy in its structure.

Density. Fabric density is also an important property of a fabric and it is expressed in terms of g/m^3. A fabric with lower density comprises a lesser amount of solid fiber phase per unit volume than a fabric with a high density (Shalev and Barker, 1983, 1984). Hence, a low density fabric can trap more air than a high density fabric. As a result, the thermal insulation characteristics of a low density fabric can be higher than a high density fabric.

Thermal conductivity. Thermal conductivity refers to the quantity of thermal energy transmitted through a unit thickness of fabric in a direction normal to the surface, with a unit temperature gradient, under steady-state conditions. If the thermal conductivity of a fabric is high, this fabric possesses lower thermal insulation. Eventually, this fabric will transmit more metabolic heat from wearers' bodies to the ambient environment and this situation can result in greater comfort to wearers. The term "thermal conductivity" of a fabric does not have a precise physical meaning because thermal energy is usually transferred through a fabric by convection, radiation, and conduction (Morris, 1953; Rees, 1941; Speakman and Chamberlain, 1930). Researchers have suggested that the appropriate term would be "thermal transmissivity" instead of "thermal conductivity" for scientific purposes. Furthermore, terms like thermal conductivity or thermal transmissivity are only applicable under thermal equilibrium conditions; in nonthermal equilibrium conditions, heat capacity needs to be evaluated to understand the thermal insulation characteristics of a fabric.

Heat capacity. Heat capacity of a fabric is the physical quantity that characterizes the amount of heat required to change a fabric's temperature by a given amount (Torvi and Threlfall, 2006). If the heat capacity of a fabric is high, more heat must be gained or lost to change the temperature of a fabric by a given number of degrees.

Consequently, a fabric with a high heat capacity should also possess good thermal insulation characteristics. Here, it is necessary to mention that heat capacity of a fabric also varies with respect to temperature and this may change the thermal insulation characteristics of the fabric. For example, the heat capacity of any synthetic fabric increases up to 50% when the temperature rises from 500 to 1000 K (Freeston, 1971).

Moisture accumulation. Some textile fibers (eg, cotton, wool) are hydrophilic in nature. Due to this, a fabric may absorb moisture (sweat) generated from wearers' bodies and accumulate it in its structure. The thermal conductivity of moisture is much higher than the thermal conductivity of air or fibers. This may cause rapid transmission of metabolic heat from wearers' bodies to the ambient environment that can result in better thermal comfort for wearers under some conditions (Barker et al., 2006; Song et al., 2011). In this context, it is necessary to remember that absorbed moisture in cotton and wool may not always increase the amount of interstitial free water inside the fabric and this situation may not cause the rapid transmission of metabolic heat from wearers' bodies. Additionally, rapid loss of metabolic heat may not always lead to thermal comfort for wearers.

3.2.4 Effect of garment properties

Although fiber, yarn, and fabric properties have significant effects on thermal insulation of clothing, many researchers have suggested that garment properties also greatly contribute to thermal insulation properties (Chen et al., 2004; Crockford et al., 1972; Song, 2007). Using different tests, they found that seams, fit, closures and location of fasteners, and microclimate regions in clothing all affect the thermal insulation characteristics of the clothing.

Stitch and seam. Clothing is usually manufactured using cut-and-sew techniques (Mandal and Abraham, 2010). This involves cutting of two-dimensional (2D) fabric panels and then sewing them together to produce a three-dimensional (3D) garment. During sewing, a thread is inserted via a needle to hold the fabric panels together. When the sewing needle moves into a fabric it may strike the assembled fibers and change the fabric structure. Due to this structural change, thermal insulation characteristics of clothing can be reduced (Glock and Kunz, 2005). Depending upon the type of stitch (eg, chain-stitch, lock-stitch, zigzag-stitch) and the type of seam (eg, plain-seam, flat-seam, lapped-seam), the thermal insulation characteristics of clothing can also vary (Mandal and Abraham, 2010).

Microclimate region. When a microclimate region becomes established between clothing and a wearer's body, this region directly contributes to the transfer of thermal energy from the wearer to the ambient environment (Ding et al., 2011a,b; Havenith et al., 1990; Keiser et al., 2008; Lawson et al., 2004; Lu et al., 2013a,b; Morozumi et al., 2012). Recently, Lu et al. (2013a) suggested that metabolic heat and sweat vapor generated on a wearer's skin may transfer through a microclimate region towards the clothing by convection, conduction, radiation, evaporation, and/or moisture absorption/diffusion (Keiser et al., 2008; Lawson et al., 2004; Lu et al., 2013b; Morozumi et al., 2012). Eventually, the microclimate region may contain a lot of moisture. This moisture may absorb thermal energy from a wearer's body. As the thermal conductivity and specific heat of water are high, sweat can enhance the transfer of metabolic heat from the wearer's body to the ambient environment. By controlling the moisture content in the microclimate region, the thermal insulation characteristics of clothing can be varied (Crockford et al., 1972; Li et al., 2012; Song, 2007).

Many researchers have tried to objectively evaluate the microclimate region. It has been usual to characterize the microclimate region mainly by calculating the average volume of the trapped air in the region; this can be readily measured using a 3D body scanning technique (Mah and Song, 2010a,b; Mandal et al., 2014; Song, 2007). In this technique, a standard sized manikin is scanned both nude and clothed, maintaining the same position and status. With a particular scanned body part (eg, chest, abdomen, thigh), the cross sections of the nude and clothed contours are subtracted and this value is considered as the air gap size for that particular body/garment combination.

Fit. Fit is an important property which affects the thermal insulation characteristics of clothing (Crown et al., 1998; Holmer and Nilsson, 1995; Lu et al., 2013a; Song, 2007). If a clothing size is too small with respect to a wearer's body, the clothing can fit tightly with the result that there is no space to establish a microclimate region between the clothing and the wearer's body (Chen et al., 2004). Alternatively, an overly loose garment can create an ineffective enclosed microclimate region between the clothing and a wearer's body and the insulation characteristics of thermal clothing can be reduced. While reduced thermal insulation can provide better comfort to wearers in some situations, reduced thermal insulation may affect the protective performance of specialized-clothing, such as thermal protective clothing or cold weather protective clothing.

Closures and location of fasteners. The closures (eg, pockets, cuffs, collar) and location of fasteners (eg, hooks, loops, buttons) are important because these help to trap air inside a garment and help to provide a microclimate between the clothing and the wearer. A large number of closures/fasteners or poorly sealed closures may disturb the microclimate environment and lower the thermal insulation of clothing (Crown et al., 1998; Lawson et al., 2004).

3.3 Measurement of thermal comfort performance of clothing

Thermal comfort performance of fabrics used in clothing is generally objectively measured using a sweating guarded hot plate apparatus. Using this apparatus, the quantities of thermal resistance, evaporative resistance, and total heat loss (THL) of fabrics can be measured and used to evaluate thermal comfort performance. In this context, the thermal resistance of a fabric can be defined as the resistance provided by the fabric to the flow of dry metabolic heat from a wearer's body to the nearby environment; the evaporative resistance of a fabric is the resistance provided by the fabric to the flow of sweat vapor from a wearer's body to the nearby environment; and, THL is the total amount of metabolic heat transferred through a fabric by the combined dry and evaporative heat exchanges (Hes and Williams, 2011; Holmer and Nilsson, 1995). A fabric with low thermal/evaporative resistance and high THL values generates lower heat stress to wearers, resulting in higher physiological comfort. In this context, a notable point is that the results obtained for a sample of fabric may not be directly applicable to any corresponding clothing without considering the amount of body surface area covered by the clothing, the distribution of the fabric and air layers on a wearer's body, the looseness or tightness of fit of the clothing, and the increase in surface area for heat loss from the clothing. In this regard, some researchers have developed models for predicting the thermal and evaporative resistances of clothing from the sweating guarded hot plate data on fabrics (McCullough et al., 1989; Mecheels and Umbach, 1976, 1977). For example, researchers at Kansas State University used sweating guarded hot plate data and clothing circumference measurements to predict the thermal and evaporative resistances of clothing layers and air layers on different parts of a body (McCullough et al., 1989). Researchers at the Hohenstein Institute of Germany conducted several studies where they used sweating guarded hot plate data to predict the intrinsic evaporative resistance of clothing ensembles, based on the fraction of the body surface area covered by each fabric (McCullough et al., 1989; Mecheels and Umbach, 1977). However, these prediction models provided no information associated with the design, fit, and construction of garments; additionally, these models showed no proven correlation with the thermal and evaporative resistances of actual clothing systems worn by people in different ambient environments (ASTM F 1868, 2014). In this context, it can be hypothesized that clothing attributes (eg, fit, design) and ambient environmental conditions should have significant effects on thermal and evaporative resistances of clothing.

Because of the preceding limitations that apply to sweating guarded hot plate tests, a great deal of research has been carried out to evaluate the thermal comfort performance of clothing using sweating thermal manikins (Meinander, 1999; Wang et al., 2011; Wu and Fan, 2009). The thermal and evaporative resistance results obtained from manikin tests can be used to indirectly predict human thermal comfort using different clothing ensembles under various ambient and metabolic conditions (Ghaddar et al., 2003; Havenith et al., 1990; ISO 7933, 2004). The advantage of these models is that predictions can be made using a wide variety of input variables, such as combinations of clothing ensembles and different ambient conditions. However, these models are based on the assumption that all human beings have uniform or constant metabolic rates. In reality, metabolic rates cannot be uniform for a group of people because of their individual differences in height, weight, and oxygen consumption; also, higher metabolic rates are observed with individuals in dynamic rather than static work situations because muscles have greater energy requirements while working. For this reason, it seems that thermal comfort of a clothed human being predicted by models based on the thermal and evaporative resistance values obtained from sweating thermal manikin tests may not always be accurate.

In response to the previously mentioned limitations of sweating thermal manikin tests, many researchers have suggested a direct approach to evaluate the thermal comfort performance of clothing, by involving human subjects (Ayogi et al., 1994; Huck and McCullough, 1988; Mountain et al., 1994; Wang et al., 2011). Here, it is noteworthy that the medical, cost, and time considerations, as well as biological variability, can also impose limitations on human trial testing, compared with sweating thermal manikin tests. However, human trials conducted in a laboratory have the advantage that data collected from real people are perceived to be, and actually can be, more reliable and valid.

The following sections describe the procedures to evaluate thermal comfort performance of clothing using the sweating guarded hot plate, sweating thermal manikin, and human trial test methods. Various internationally recognized standards to evaluate the thermal comfort performance are discussed, and the limitations of these standards are stated. This section may help to improve the existing standards or develop new standards to evaluate thermal comfort performance.

3.3.1 Sweating guarded hot plate tests

For evaluating the thermal and evaporative resistances of fabrics using a sweating guarded hot plate, the ISO 11092 standard was developed in 1993 by scientists at the Hohenstein Institute in Germany (ISO 11092, 2014). Then, the NFPA (National Fire Protection Association) developed a method to determine THL using thermal protective fabrics or fabric systems using the ISO 11092 standard. Later, this THL evaluation method was added to several NFPA standards—NFPA 1971, NFPA 1977, NFPA 1951, and NFPA 1999 (Godhlke, 1997; NFPA 1971, 2007; NFPA 1977, 2011; NFPA 1951, 2013; NFPA 1999, 2013). Next, the members of the ASTM F23 committee decided to compile the evaluation procedures for measuring thermal resistance, evaporative resistance, and THL in one document, which resulted in the ASTM F 1868 standard (ASTM F 1868, 2014; Godhlke, 1997; Huang, 2006; McCullough et al., 2004; Wang et al., 2011). By using these standard methods, researchers have tried to objectively correlate the thermal/evaporative resistance and/or THL values of fabrics with the thermal comfort of human beings. For example, the Hohenstein Institute in Germany has suggested that a rain suit fabric with evaporative resistance of $<6\,m^2$ Pa/W provides very good comfort to wearers; similarly, using the European standard EN 469, it has been suggested that the evaporative resistance of fabrics used in firefighters' protective clothing should be less than $30\,m^2$ Pa/W (EN 469, 2005); also, using the NFPA 1971:2007 standard method, it has been suggested that the THL of fabric used in thermal protective clothing should be at least $205\,W/m^2$ in order to provide physiological comfort to working firefighters (NFPA 1971, 2007).

The ASTM F 1868 standard method is widely used to evaluate/calculate thermal resistance, evaporative resistance, and THL of fabrics (ASTM F 1868, 2014). However, this standard possesses several limitations. For example, reaching a steady-state condition with some fabrics is often quite difficult during the evaluation of thermal and evaporative resistances; the fabric samples may lift off from the plate, especially when tested under an ambient air velocity of 1 m/s, and this can result in abnormally high thermal and evaporative resistance values (resulting in low THL values). The standard does not specify the direction of the ambient air flow or the level of air turbulence to evaluate the thermal and evaporative resistance; and the overall THL calculation procedure is very tedious and time consuming. Furthermore, although the ASTM F 1868 standard is the best way to isolate and evaluate thermal resistance and evaporative resistance individually, some researchers feel that it is not realistic, as dry and evaporative heat losses occur simultaneously from a human body to the ambient environment, and there may be interactions between the dry and evaporative heat flows which this test does not capture (McCullough et al., 2004). Additionally, the devices used in ASTM F 1868 for determining clothing properties ignore the physiological state of the wearer and are inadequate to evaluate the transient thermal properties of clothing. Keeping this in mind, Psikuta et al. (2013) have developed a thermo-physiological human simulator that can provide more realistic comfort property ratings of clothing. By using this human simulator, ISO is currently developing a standard (ISO CD 18640-1: Protective clothing for firefighters- physiological impact—Part 1: Measurement of coupled heat and mass transfer with the sweating TORSO) that can be used to evaluate the physiological comfort properties of clothing.

3.3.2 Sweating thermal manikin tests

For evaluating the thermal and evaporative resistances of clothing using full-scale manikins, an ISO 9920 standard was developed in 1995 and later modified in 2007. In 2010, ASTM F 1291 and ASTM F 2370 standards were developed and later modified to evaluate the thermal and evaporative resistances of clothing using full-scale manikins, respectively (ASTM F 1291, 2010; ASTM F 2370, 2010; Fukazawa et al., 2004; ISO 9920, 2007; Wang et al., 2011; Wu and Fan, 2009). It is notable that a few other standards are also available to evaluate thermal resistance, ISO 15831, EN 342, and ASTM F 1720, along with the ASTM F 1291 standard (ASTM F 1720, 2014; ASTM F 1291, 2010; EN 342, 2004; ISO 15831, 2004). Although the ISO 15831, EN 342, and ASTM F 1720 standards use the same measurement technique for thermal resistance evaluation as described in the ASTM F 1291 standard, these four standards have different scopes associated with (1) manikin features, (2) test conditions, (3) methods to calculate the thermal resistance, and (4) parameters for test results (Huang, 2006). In both of the ASTM F 1291 and ASTM F 2370 standards, a standing manikin built in the shape and size of an adult male or female is dressed with a garment. By controlling the ambient environmental conditions (temperature, air velocity) and/or the simulation of sweating by the manikin, the thermal and evaporative resistances of a garment can be measured according to the methods described in these standards (ASTM F 1291, 2010; ASTM F 2370, 2010). A notable point is that the thermal and evaporative resistance values of clothing in the ASTM F 1291 and ASTM F 2370 standards are evaluated in a normal ambient environment (relatively calm and cooler than the manikin body); thus, these methods may not be accurately applicable to evaluate the thermal comfort performance of specialized clothing, where wearers are exposed to extreme thermal environments.

The ASTM F 1291 and ASTM F 2370 standards are widely used to quantify and compare thermal and evaporative resistance values provided by clothing ensembles having different designs, fabrics, garment layers, closures, and fits (ASTM F 1291, 2010; ASTM F 2370, 2010); the thermal and evaporative resistance values of clothing ensembles obtained from these standards can further be used in models to empirically predict the physiological responses of people in different environmental conditions

(Ghaddar et al., 2003; Havenith et al., 1990). However, the ASTM F 1291 and ASTM F 2370 standards possess several limitations. For example, these are only static tests that provide a baseline clothing measurement on a standing manikin, hence, the effects of body positions and movement are not addressed; the thermal and evaporative resistance values obtained are applicable only to the particular clothing evaluated and specified environmental conditions of each test, particularly with respect to ambient air velocity and/or sweating simulations; the measurement of thermal and evaporative resistances provided by clothing ensembles is a complex process and depends on the apparatus and technique used; technical knowledge concerning the theory of heat transfer, temperature, air motion measurement, and testing practices, are essential for an operator to evaluate thermal and evaporative resistances. Additionally, the ASTM F 2370 standard has been developed based on the premise that the vapor pressure at the manikin's skin surface is 100%, and the sweating rate should be high enough to saturate the skin. However, this premise is not true in unstaged scenarios, in which optimal sweating rates of human beings may vary due to many factors, such as the ambient environmental conditions, activities performed, and type of clothing worn. It has been observed that if the sweat rate is low and the clothing is thin, all of the water vapor dissipates through the clothing towards the ambient environment; in this situation, 100% water vapor pressure at the skin surface cannot be attained. It seems that the ASTM F 2370 standard should measure the exact water vapor pressure near the skin surface, although this is very difficult. Furthermore, it is notable that ASTM F 1291 and ASTM F 2370 standards designate a test protocol for determining the thermal and evaporative resistance of the clothing; however, these standards do not specify anything about the posture of the manikin. Additionally, ASTM F 2370 standard does not stipulate a specific design for sweating simulation and thus, leaves the sweating mechanism open to interpretation (ASTM F 1291, 2010; ASTM F 2370, 2010). Interestingly, this unspecified freedom has encouraged the development and use of manikins in different laboratories with different sweating systems, sweat rate evaluation techniques, movement techniques, etc. (Fan and Chen, 2002; Meinander, 1999). Some of these manikins are: (1) the Finnish sweating thermal manikin "Coppelius", (2) the Japanese sweating thermal manikin "Taro", (3) the Swiss sweating thermal manikin "SAM", (4) the Hong Kong sweating thermal manikin "Walter", (5) the United States sweating thermal manikin "ADAM", (6) the United States sweating thermal manikin "Newton", and (7) the Japanese sweating thermal manikin "KEM" (Dozen et al., 1992; Fan and Chen, 2002; Fukazawa et al., 2004; Meinander, 1999; Rehn, 2004; Richards and Mattle, 2001). Recently, working mechanisms of each of these widely used manikins have been thoroughly discussed by Mandal et al. (in press).

3.3.3 Human trials

The ASTM F 2668 standard is available to evaluate thermal comfort performance of clothing using human trials in a controlled laboratory environment (ASTM F 2668, 2011); however, the actual conditions are at the investigators' discretion. Additionally, field trials can be carried out according to investigators' requirements and objectives. The ASTM F 2668 standard is believed to be appropriate for the evaluation of the majority of specialized-clothing ensembles, especially where wearers need to walk or perform similar activities; thus, this standard utilizes a treadmill for the wearers' exercise protocol. In certain situations, when the user is required to wear a specialized-clothing ensemble while performing functions such as sitting or standing with only arm movement, alternative exercise equipment (eg, arm cycle-ergonometers) or protocols should be considered in determining thermal comfort performance of clothing. According to the ASTM F 2668 standard, a group of healthy and medically fit human subjects (selected based on a screening test) are required to wear the clothing that needs to be tested, and then, the clothed wearers are required to run on a treadmill (at a predetermined speed and angle of inclination) in a particular ambient environment (a fixed temperature and air velocity). While running, wearers' physiological parameters, such as the maximum heart rate responses, maximum oxygen consumption, core body temperatures (esophageal, rectal, and intestinal), and mean skin temperatures (forehead, right scapula, left upper chest, upper right arm, lower left arm, left hand, right anterior thigh, and left calf) are measured after a fixed interval of time using a metabolic measurement cart, thermistors, heart rate meter, etc.; core and skin temperatures can be measured according to the ISO 9886 standard (ISO 9886, 2004). The "whole body sweat rate" of the participants is determined by subtracting the measured postdry unclothed body mass (if urination occurs during testing, the mass of urine shall be added to the postbody mass; if fluid consumption occurs during the testing, the mass of the fluid must be subtracted from the postbody mass) from predry unclothed body mass; and, this rate can be used to assess the level of dehydration caused by the thermal stress under the test conditions. Alternatively, subjective evaluations of the participants' comfort can be recorded after certain intervals using the Likert or a rating scale (Likert, 1932); in order to avoid any influence on their subjective ratings, the participants should not be informed about the details of the study, such as selection of clothing and ambient environments (eg, temperature, air velocity). The termination of the test should result either by voluntary withdrawal, signs of impending heat stress (eg, disorientation, chills, or nausea), if the heart rate reaches 90% of the measured maximum during baseline assessment, or if core temperature exceeds 39°C, or a skin temperature measurement at any location exceeds 38°C. All participants must attempt to last as long as possible to allow the physiological response to become sufficiently evident; a maximum time of 2h should be allotted for this test. A recovery period is provided for participants upon completion of each test condition; at the end of the recovery period, the core temperature and heart rate should be below 38°C and 100 beats/min, respectively.

In human trial tests, careful consideration should be given regarding appropriate test methodology, as the results obtained from the trials are often used to provide garment users with information from a more realistic perspective. Since even homogeneous groups of human test subjects display considerable interindividual variability, every effort must be made to account for variables within the investigators' control in order to collect meaningful thermal comfort data. Investigators must also adhere to appropriate regulations for protection of human research subjects and comply with the recommendations of their institutional ethics review boards. Also, although human trials are best done in controlled laboratory environments, human trials can also be conducted in actual working field environments for realism and clothing user acceptability. Here, it is extremely important to remember that it is ethically unacceptable to involve human subjects in the evaluation of thermal comfort performance of specialized clothing in hazardous thermal environments that wearers (eg, firefighters, industrial workers) actually face while on duty. Thus, researchers should use a temperate environmental chamber to evaluate the comfort provided by specialized clothing. During all human trials in a laboratory-controlled environment, a physician should be readily available; for human trials in a field environment, an ambulance service should be readily available in addition to the physician. In summary, it can be concluded that extensive precautions, financial investment, and test designs are necessary, during and after the trials, because of the duty of care owed to human subjects involved in this type of test. Additionally, ethical issues are involved in recruitment of research participants, especially the consent to exercise to exhaustion. This is the reason that very few studies are available on the evaluation of thermal comfort provided by clothing using human trials (Huck and McCullough, 1988; Mountain et al., 1994; Semenniuk et al., 2005; Wang et al., 2011).

3.4 Critical assessment of thermal comfort performance of clothing

Black and Matthew (1934) and Rees (1946) examined the effect of environmental air relative humidity on the thermal resistance of fabrics. It was found that thermal resistance was highly dependent upon relative humidity. They concluded that fabrics gain different levels of moisture under different levels of humidity, and this gained moisture affects thermal resistance. Black and Matthew (1934) proved through experimentation that a marked reduction in thermal resistance takes place when the moisture content increases from 0% to 75% of the dry weight of fabrics. Barker and Heniford (2011) evaluated the thermal resistance of various inherently fire-resistant woven/nonwoven fabrics or multilayered fabric systems. In these studies, it has been found that fabric weight and thickness definitely affect thermal resistance; however, some other fabric features (eg, air permeability, porosity, surface area) were also equally important for thermal resistance (Ding et al., 2011a,b). Shekar et al. (2009) found that environmental air velocity may cause thermal resistance loss in a fabric system; however, the presence of an impervious outer layer in the fabric system helps to reduce the loss in thermal resistance under high wind velocity conditions. They also found that the thermal resistance of a fabric system is independent from the nature of the outer layer (pervious or impervious), under normal environmental conditions. Additionally, it was determined that nonwoven fabrics possessed high thermal resistance due to their bulkiness, compression recovery, and porosity compared to woven fabrics (Song et al., 2011; Sun et al., 2000). However, wet nonwoven fabrics possessed lower thermal resistance than dry nonwoven fabrics; this was because the thermal conductivity of the wet fabric was much higher than that of the dry fabric. Barker and Heniford (2011) stated that the construction and thickness of fire-resistant nonwoven fabrics mainly affect thermal resistance. This study demonstrated that the layering of fiber webs in nonwoven fabrics was a potential means to enhance thermal resistance because it contributed air layers and thickness without any increase in weight of the nonwoven fabrics. Wu et al. (2011) evaluated the thermal resistance of various types of clothing using different postures (standing, sedentary, supine) of the sweating thermal manikin "Walter". In this study, it was evident that the thermal resistance of clothing was significantly higher in the sedentary posture than in the standing posture. Here, the radiative heat transfer coefficient from the manikin body to the ambient environment was lower in the sedentary posture than in the standing posture, due to a reduction in the radiative body surface area in the sedentary posture; this lower radiative heat transfer coefficient increased the thermal resistance of the clothing. The sedentary posture also created a cavity over the horizontal knees and thighs of the manikin. As a consequence, natural convection reduced over the manikin body in the sedentary posture versus the standing posture; this reduction in natural convection enhanced the thermal resistance of the clothing. Furthermore, it was evident that thermal resistance was significantly higher in the supine posture than in the standing posture. This is because the manikin remained flat on a wooden bed in the supine posture, and this wooden bed enhanced thermal resistance. In this study, it was evident that thicker clothing always had a higher thermal resistance than thinner clothing in all postures. However, the thickness of the clothing may have been significantly reduced in the supine posture due to compression provided by the manikin body; this reduction in thickness significantly reduced the thermal resistance of the clothing.

Gibson (1993) explored the evaporative resistance of various woven and nonwoven fabric materials. These materials included permeable and impermeable types tested as single layers, laminates, and composites. It was found that the evaporative resistance of permeable materials was very low, while the evaporative resistance of impermeable materials was significantly higher. This is because impermeable materials do not allow the transfer of moisture vapor through their structure, whereas permeable materials allow moisture vapor-transfer through their structure at a high rate. Also, the evaporative resistances of permeable materials were evaluated under a variety of conditions: (1) using varying directions and velocity of airflow over the materials, and (2) by providing an air gap between the material sample and the sweating skin stimulating hot plate. It has been found that airflow conditions had a significant effect on evaporative resistance, and the open structure of the material became particularly important for evaporative resistance, especially when an air gap existed between the material sample and the sweating skin stimulating hot plate. Gibson (1993) concluded that the correlation between the open structure and evaporative resistance can be altered by varying the thickness of the materials, at an air velocity of 1–2 m/s. Wu et al. (2011) evaluated the evaporative resistance of various types of clothing worn by the sweating thermal manikin "Walter" in different postures (standing, sedentary, supine). They identified a high correlation between the evaporative resistance measured both in standing and sedentary postures. It was observed that the evaporative resistance in the sedentary posture was about 20–97% higher than that measured in the standing posture. In this context, Havenith et al. (1990) mentioned that evaporative resistance measured in a sedentary posture is about 16–38% higher than that of the standing posture. It seems that Wu and Havenith's studies provided different ranges of values of the difference in evaporative resistance between the sedentary and standing postures; nevertheless, Wu's study may be more realistic because they used the sweating thermal fabric manikin "Walter", a highly reliable testing method. Wu et al. (2011) also stated that the evaporative resistance in the manikin's supine posture was significantly higher than that in the standing or sedentary posture. This is because a large amount of water condensation occurred (within the clothing, and in between the clothing and the bed, etc.) in the supine posture, and this water condensation enhanced evaporative resistance. In this study, it was found that evaporative resistance under nonisothermal conditions was generally lower than under isothermal conditions, in both standing and sedentary postures. This is because the natural convection induced by the temperature gradient was much lower under isothermal conditions than with nonisothermal conditions. It was also observed that a higher temperature gradient in between the manikin body and the ambient environment contributed to accumulation of water within the manikin's clothing; this situation drastically changed the evaporative resistance. In fact, the moisture accumulation in the isothermal condition was around 1%, or close to zero, which was substantially lower than that under the nonisothermal condition. In the nonisothermal condition, some moisture vapor accumulated in the clothing ensemble instead of fully evaporating. The evaporative resistance measured in the nonisothermal condition, therefore, was apparently substantially lower than that under the isothermal condition.

Many researchers have corroborated that heat loss through fabrics may occur through combined heat and moisture/water vapor transfer by conduction, convection, radiation, evaporation, and/or diffusion (Farnworth, 1986; Ghaddar et al., 2003; Song et al., 2011; Tian et al., 2012; Wang et al., 2011). Farnworth (1986) studied heat loss by modeling the combined heat and water vapor transfer through multilayered fabrics. Heat transfer occurred by conduction and radiation; whereas, water vapor transfer occurred by diffusion. In this study, it was evident that hygroscopic and nonhygroscopic fabrics transfer heat and water vapor in different ways. Thicker fabrics did not allow the transfer of heat and water vapor through their structures, which ultimately reduced heat losses. Additionally, hygroscopic fabrics absorb water vapor and transfer it to the ambient environment; this phenomenon ultimately enhanced heat losses through the fabrics. It was also found that water-impermeable but vapor-permeable fabrics possessed an excellent heat loss characteristic at high ambient temperatures; however, this characteristic was not prominent at low ambient temperatures. Farnworth (1986) concluded that a layered fabric may not allow the transfer of heat and water vapor through its structure, which ultimately can reduce heat loss. Recently, Tian et al. (2012) analyzed the heat loss behavior of multilayered fabric systems used in specialized clothing. The multilayered fabric systems were composed of three different fabrics in this study and the fabrics were arranged in different layering sequences. Altogether, six different three-layered fabric systems were prepared using different layer stacking sequences. In this study, the stacking sequence of the three-layered fabric systems was found to play an important role in heat loss. The volumetric heat capacity of the layer contacting the heat source was the prime parameter that determined the heat loss through the three-layered fabric systems.

Wu et al. (2009) studied the thermal comfort provided by different types of T-shirts (made of ten kinds of hygroscopic fibers: cotton, wool, lyocell, modal, soybean, bamboo, and their blends) using human trials. The experimental results showed that the thermal wet comfort of the T-shirts varied due to the fiber types, and was mainly influenced by the heat and moisture transfer during exercise. It was found that the natural hygroscopic fibers (eg, cotton and wool) were damper and provided greater thermal comfort than the other fibers (eg, regenerated cellulosic fibers, bamboo, soybean, and modal). Additionally, modal fiber was found to be particularly adhesive to wearers' bodies; bamboo had a significant coolness to the touch; and lyocell showed a higher thermal sensation than the other regenerated cellulosic fibers, with moderate dampness and stickiness. The comfort sensations of different body parts of the wearers were also found to be very different during exercise. It seems that there is a need to design T-shirts to provide better comfort to wearers. Hostler et al. (2010) compared the comfort provided by thermal protective clothing and regular clothing based on physiological parameters (eg, hydration and maximum oxygen consumption) of firefighters during volitional exhaustion tests. It was established that thermal protective clothing possesses lower comfort than regular-clothing because it causes dehydration to firefighters. This dehydration means that firefighters quickly reach their maximum oxygen consumption levels. This situation results in high heat stress for firefighters during their work.

Overall, it can be concluded that previous studies have identified that fabric features (eg, fiber types, weaves, design, weight, thickness, porosity), clothing attributes (eg, fit, design, construction), and/or ambient environmental variables (air, temperature, relative humidity) mainly affect the metabolic-heat and/or sweat-vapor transfer (convective/conductive/radiative/diffusive) through clothing. This situation ultimately affects the thermal resistance, evaporative resistance, and/or THL of the clothing. In this context it is necessary to remember that it is very difficult to obtain a clear relationship between the individual fabric-features/clothing-attributes and thermal resistance, evaporative resistance, and/or THL. This is because most of these features/attributes are so profoundly interrelated that it is impossible to separate them.

3.5 Key issues related to thermal comfort of clothing

Some of the key issues related to thermal comfort performance evaluation and development of new fabrics are discussed in this section. These key issues need to be thoroughly researched and resolved in order to develop high-performance and cost-efficient clothing that can provide better comfort to wearers.

3.5.1 Development of state-of-the-art testing methods

Based on the preceding discussion, the ISO 9920, ASTM F 1291, and ASTM F 2370 standards designate test protocols for determining the thermal comfort performance (thermal and evaporative resistance) of clothing using sweating thermal manikins. However, these standards do not specify anything about the postures of the manikins.

Additionally, ISO 9920 and ASTM F 2370 standards do not stipulate a specific design for sweating simulation and also leave the sweating mechanism open to interpretation (ASTM F 1291, 2010; ASTM F 2370, 2010). In order to realistically evaluate the thermal and evaporative resistances of clothing, it is expected that these standard methods should be improved by clearly specifying posture and sweating aspects of manikins. Furthermore, sweat generation rates of presently available sweating thermal manikins (Coppelius, Taro, SAM, Walter, Newton, ADAM, KEM) are very limited compared to the sweat generated by some wearers (eg, firefighters, mountaineers) under their actual conditions of work. Thus, the evaluation of evaporative resistance of specialized clothing using sweating thermal manikins may not relate to real situations. It is expected that the sweat generation rates of the presently available manikins should be increased in order to accurately simulate wearers' physiological aspects when wearing specialized clothing.

Furthermore, thermal comfort performance (thermal resistance, evaporative resistance, and/or THL) of fabrics/clothing is mainly evaluated using tests based on standard sweating guarded hot plates, sweating thermal manikins, and human trial tests (ASTM F 1868, 2014; ASTM F 1291, 2010; ASTM F 2370, 2010; ASTM F 2668, 2011). However, these tests are very expensive, cumbersome, and require skilled technicians to do the testing. Keeping this in mind, many researchers resort to numerical modeling to simulate the thermal and evaporative resistances of fabrics/clothing (Fan and Chen, 2002; Fan and Cheng, 2005; Farnworth, 1986; Li and Holcombe, 1998; Ogniewicz and Tien, 1981; Wu and Fan, 2009). These studies yield information about the mathematical interactions between fabrics' attributes and thermal/evaporative resistances. However, the practical applications of these mathematical models for calculating thermal and evaporative resistances may be limited due to their intrinsic complexities. It is suggested that the development of cost-efficient and user-friendly empirical models (eg, multiple linear regression, artificial neural network) by employing the fabric and clothing properties is necessary to efficiently calculate the thermal comfort performance of clothing (Mandal and Song, 2014).

3.5.2 Development of new fabrics

Although many innovative approaches have been applied to develop different types of clothing, specialized clothing may still cause significant heat stress for wearers by exerting high loads on them and/or improperly dissipating their metabolic heat and sweat vapor. In specialized clothing, the presence of (1) heavy and thick thermal liners (for providing insulation against heat transfer), and/or (2) water-impermeable and partially breathable moisture barriers (for providing barriers to mass transfer) cause significant heat stress to wearers' bodies (Bernard et al., 2010; Sun et al., 2000). Although the use of thermal liners and/or moisture barriers is inevitable in order to provide wearers with protection against occupational hazards (eg, flame, radiant heat, steam), modern design philosophy demands the development of specialized clothing that can provide optimal thermal comfort to wearers by reducing their heat stress. Keeping this concern in mind, researchers have developed new fabrics that can provide both good thermal comfort and protection to wearers. For example, Hocke et al. (2000) developed a nontextile thermal liner by using foamed silicon on a vapor-permeable moisture barrier, which can enhance wearers' thermal comfort and protection; Holme (2004) macroencapsulated nanoporous gels in viscose-based nonwoven fabric to produce a lightweight thermal liner that was able to provide adequate thermal comfort to wearers; and Jin et al. (2011) padded a needle-punched nonwoven thermal linear with 5% on an aerogel dispersed in acetone. As the aerogel comprised extremely fine nanopores with 99.8% air, they enhanced the insulation features of the thermal liner without making it heavier (Song and Lu, 2013). Similarly, the application of nanoclay reinforced resin coatings in thermal liners was found to enhance their insulation features without compromising their weight. Dadi (2010) also suggested that the application of lightweight nanofiber-based fabrics, nanofinishes and/or phase change materials (PCMs) on traditional fibers/fabrics can reduce the heat stress for wearers (Carter et al., 2007; Chou et al., 2008). Here, PCMs can be applied during either fiber spinning or chemical finishing (coating, lamination) processes. In this context, the protective performance of specialized clothing, with or without PCMs, under radiant heat exposures was studied using a theoretical heat transfer model and it was found that application of PCMs in specialized clothing would provide equivalent protective performance with reduced garment thickness/weight (McCarthy and Marzo, 2011). Similarly, the UK Defense Clothing and Textiles Agency demonstrated that the application of shape memory materials (SMMs) can also enhance the protective and comfort performance of certain types of specialized clothing. This is because SMM (alloy, polymers) can change their current shape into a prescribed crystal structure shape depending upon thermal environment (Otsuka and Wayman, 1998). For example, an application involving shape memory nickel-titanium alloy finishes on a multilayered fabric may widen the air gap between layers in the presence of high temperatures. This increased air gap can help to provide better thermal protection for wearers from extreme thermal hazards without adding any heat stress to them (Lee et al., 2007). Furthermore, a group of researchers has suggested further development of high-performance fabrics that can passively control wearers' thermal comfort and protection (Chou et al., 2008). They recommended applying cooling resources (eg, ice, frozen gels) to specialized clothing in order to actively alleviate heat stress caused by the clothing. Chitrphiromsri et al. (2006) developed a type of intelligent specialized clothing in which liquid water was injected into the outer surface of the clothing, and the injection process was activated by temperature sensors embedded in the clothing. The water injected absorbed a large amount of thermal energy and this evaporated in the presence of thermal hazards. Hence, this mechanism can actively enhance the performance of clothing by providing greater burn protection and thermal comfort to wearers. Additionally, the application of emerging technologies in specialized clothing can also passively control wearers' thermal comfort by monitoring their work situations (Donnelly et al., 2006; Song and Lu, 2013). For example, weaving of wires and sensors into fabrics can provide information to wearers about the hazards they are encountering. There are other benefits as well: integration of thermal sensors into the interior and exterior of specialized clothing can be used to control the temperature near wearers, inside the clothing, and close to wearers' bodies. Incorporation of a circularly polarized antenna (with a bandwidth of more than 180 MHz), linked to a two-way portable communication device within clothing can be useful to monitor wearers' working situations. The addition of a gas detector into specialized clothing can monitor the presence of life threatening hot gasses in the ambient environment of wearers; and, the attachment of a personal alert safety system in the clothing can trigger an alarm when a wearer may experience extreme temperature hot or cold situations.

Many developments can provide optimal thermal comfort and protection to wearers. However, many of these developments are not yet cost effective and are confined to laboratory settings and/or are only applied in highly specialized circumstances (eg, aerospace, military, defense). Development of these technologies in the mass markets should be aided by cost-efficient and revolutionary applications of new nanofibers and nanofinishes, and smart textiles based on PCMs or SMM, cooling devices (Gibson et al., 2007; Grafe and Graham, 2002). As nanofibers are lighter and thinner than conventional macrofibers, it can be hypothesized that the use of nanofibers can reduce the weight and thickness of clothing and may be used to significantly improve the thermal comfort of wearers. Additionally, smart textiles that can sense wearers' physiological conditions, postures, and ambient environment can provide promising developments for the mass-production of specialized comfortable clothing (Shin et al., 2005).

References

ASTM F 1291, 2010. Standard test method for measuring the thermal insulation of clothing using a heated manikin.

ASTM F 1720, 2014. Standard test method for measuring thermal insulation of sleeping bags using a heated manikin.

ASTM F 1868, 2014. Standard test method for thermal and evaporative resistance of clothing materials using a sweating hot plate.

ASTM F 2370, 2010. Standard test method for measuring the evaporative resistance of clothing using a sweating manikin.

ASTM F 2668, 2011. Standard practice for determining the physiological responses of the wearer to protective clothing ensembles.

Ayogi, Y., McLellan, T.M., Shephard, R.J., 1994. Effects of training on heat tolerance in exercising men wearing protective clothing. Eur. J. Appl. Physiol. Occup. Physiol. 68 (3), 234–245.

Barker, R.L., 2008. Multilevel approach to evaluating the comfort of functional clothing. J. Fiber Bioeng. Inf. 1 (3), 173–176.

Barker, R.L., Heniford, R.C., 2011. Factors affecting the thermal insulation and abrasion resistance of heat resistant hydro-entangled nonwoven batting materials for use in firefighter turnout suit thermal liner systems. J. Eng. Fibers Fabr. 6 (1), 1–10.

Barker, R.L., Guerth-Schacher, C., Grimes, R.V., Hamouda, H., 2006. Effects of moisture on the thermal protective performance of firefighter protective clothing in low level radiant heat exposures. Text. Res. J. 76 (1), 27–31.

Behnke, W.P., 1984. Predicting flash fire protection of clothing from laboratory tests using second-degree burn to rate performance. Fire Mater. 8 (2), 57–63.

Bernard, T., Ashley, C., Trentacosta, J., Kapur, V., Tew, S., 2010. Critical heat stress evaluation of clothing ensembles with different levels of porosity. Ergonomics 53 (8), 1048–1058.

Bhattacharjee, D., Kothari, V.K., 2009. Heat transfer through woven fabrics. Int. J. Heat Mass Transf. 52 (7–8), 2155–2160.

Black, C.P., Matthew, J.A., 1934. The physical properties of fabrics in relation to clothing. Part II—water vapour permeability of fabrics. J. Text. Inst. 25 (7), T225–T240.
Bogaty, H., Hollies, N.R.S., Harris, M., 1957. Some thermal properties of fabrics. Part I: the effect of fiber arrangement. Text. Res. J. 27 (6), 445–449.
Bozkurt, E., Kaya, E., Tanoglu, M., 2007. Mechanical and thermal behavior of non-crimp glass fiber reinforced layered clay/epoxy nanocomposites. Compos. Sci. Technol. 67 (15–16), 3394–3403.
Carter, J.M., Rayson, M.P., Wilkinson, D.M., Richmond, V., Blacker, S., 2007. Strategies to combat heat strain during and after firefighting. J. Therm. Biol. 32 (2), 109–116.
Celcar, D., Meinander, H., Gersak, J., 2008. A study of the influence of different clothing materials on heat and moisture transmission through clothing materials, evaluate using sweating cylinder. Int. J. Clothing Sci. Technol. 20 (2), 119–130.
Chen, Y.S., Fan, J.T., Qian, X., Zhang, W., 2004. Effect of garment fit on clothing thermal insulation. Text. Res. J. 74 (8), 742–748.
Chitrphiromsri, P., Kuznetsov, A.V., Song, G., Barker, R.L., 2006. Investigation of feasibility of developing intelligent firefighter-protective garments based on the utilization of a water-injection system. Numer. Heat Transfer Part A Appl. 49 (5), 427–450.
Chou, C., Tochihara, Y., Kim, T., 2008. Physiological and subjective responses to cooling devices on firefighting protective clothing. Eur. J. Appl. Physiol. 104 (2), 369–374.
Copper, J.A., Nguyen, D.D., Ruby, B.C., Schoeller, D.A., 2011. Maximal sustained levels of energy expenditure in humans during exercise. Med. Sci. Sports Exerc. 43 (12), 2359–2367.
Crockford, G.W., Crowder, M., Prestidge, S.P., 1972. A trace gas technique for measuring clothing microclimate air exchange rates. Br. J. Ind. Med. 29 (4), 378–386.
Crown, E.M., Ackerman, M.Y., Dale, J.D., Tan, Y., 1998. Design and evaluation of thermal protective flightsuits. Part II: instrumented mannequin evaluation. Clothing Text. Res. J. 16 (2), 79–87.
Dadi, H.H., 2010. Literature Overview of Smart Textiles (M.Sc. thesis). University of Borås, Borås.
Ding, D., Tang, T., Song, G., McDolnald, A., 2011a. Characterizing the performance of a single-layer fabric system through a heat and mass transfer model. Part I: heat and mass transfer model. Text. Res. J. 81 (4), 398–411.
Ding, D., Tang, T., Song, G., McDolnald, A., 2011b. Characterizing the performance of a single-layer fabric system through a heat and mass transfer model. Part II: thermal and evaporative resistance. Text. Res. J. 81 (9), 945–958.
Donnelly, M.K., Davis, W.D., Lawson, J.R., Selepak, M.J., 2006. Thermal Environment for Electronic Equipment Used by First Responders, National Institute of Standards and Technology—Technical Note 1474, National Institute of Standards and Technology, USA, 1–36.
Dozen, Y., Aratani, Y., Saitoh, T., Tsuchida, K., Harada, K., Takenishi, S., 1992. Modeling of sweating manikin. J. Text. Mach. Soc. Jpn. 38 (4), 101–112.
EN 342, 2004. Protection against cold.
EN 469, 2005. Protective clothing for firefighters: performance requirements for protective clothing for firefighting.
Fan, J.T., Chen, Y.S., 2002. Measurement of clothing thermal insulation and moisture vapor resistance using a novel perspiring fabric thermal manikin. Meas. Sci. Technol. 13 (2), 1115–1123.
Fan, J., Cheng, X.Y., 2005. Heat and moisture transfer with sorption and phase change through clothing assemblies. Part II: theoretical modeling, simulation, and comparison with experimental results. Text. Res. J. 75 (3), 187–196.

Fan, J.T., Luo, Z., Li, Y., 2000. Heat and moisture transfer with sorption and condensation in porous clothing assemblies and numerical simulation. Int. J. Heat Mass Transf. 43 (12), 2989–3000.

Farnworth, B., 1983. Mechanism of heat flow through clothing insulation. Text. Res. J. 53 (12), 717–725.

Farnworth, B., 1986. A numerical model of the combined diffusion of heat and water vapour through clothing. Text. Res. J. 56 (11), 653–665.

Fourt, L., Hollies, N.R.S., 1970. Clothing: Comfort and Function. Martin Dekker, New York.

Freeston, W.D., 1971. Flammability and heat transfer of cotton, Nomex, and PBI fabric. J. Fire Flammab. 2 (1), 57–76.

Frydrych, I., Dziworska, G., Bilska, J., 2002. Comparative analysis of the thermal insulation properties of the fabrics made of natural and man-made cellulose fibres. Fibres Text. East. Eur. 10 (4), 40–44.

Fukazawa, T., Lee, G., Matsuoka, T., Kano, K., Tochihara, Y., 2004. Heat and water vapor transfer of protective clothing system in a cold environment, measured with a newly developed sweating thermal manikin. Eur. J. Appl. Physiol. 92 (6), 645–648.

Ghaddar, N., Ghali, K., Jones, B., 2003. Integrated human-clothing system model for estimating the effect of walking on clothing insulation. Int. J. Therm. Sci. 42 (6), 605–619.

Ghassemi, A., Mojtahedi, M.R.M., Rahbar, R.S., 2011. Investigation on the physical and structural properties of melt-spun multifilament yarns, drawn yarns and textured Yarns produced from blend of PP and oxidized PP. Fibers Polym. 12 (6), 789–794.

Gibson, P.W., 1993. Factors influencing steady-state heat and water vapour transfer measurement for clothing material. Text. Res. J. 63 (12), 749–764.

Gibson, P.W., Lee, C., Ko, F., Reneker, D., 2007. Application of nanotechnology to nonwoven thermal insulation. J. Eng. Fibers Fabr. 2 (2), 32–40.

Gilligan, I., 2010. The prehistoric development of clothing: archaeological implications of a thermal model. J. Archeol. Method Theory 17 (1), 15–80.

Glock, R.E., Kunz, G.I., 2005. Apparel Manufacturing: Sewn Product Analysis. Pearson Prentice Hall, Upper Saddle River, NJ.

Godhlke, D.J., 1997. History of the development of the total heat loss test method. In: Stull, J.O., Schwope, A.D. (Eds.), Performance of Protective Clothing: 6th Volume. ASTM International, Orlando, FL, pp. 109–226.

Grafe, T., Graham, K., 2002. Polymeric nanofibers and nanofiber webs: a new class of nonwovens. In: International Nonwovens Technical Conference, Atlanta, USA.

Greenwood, K., Rees, W.H., Lord, J., 1970. Studies in Modern Fabric. The Textile Institute, Manchester.

Hatch, K.L., 1993. Textile Science. West Publishing Company, New York.

Havenith, G., Heus, R., Lotens, W.A., 1990. Clothing ventilation, vapour resistance and permeability index: changes due to posture, movements and wind. Ergonomics 33 (8), 989–1005.

Hes, L., Williams, J., 2011. Laboratory measurement of thermophysiological Comfort. In: Song, G. (Ed.), Improving Comfort in Clothing. Woodhead Publishing Limited, Cambridge, pp. 114–137.

Hes, L., Araujo, M., Djulay, V., 1996. Effect of mutual bonding of textile layers on thermal insulation and thermal-contact properties of fabric assemblies. Text. Res. J. 66, 245–250.

Hocke, M., Strauss, L., Nocker, W., 2000. Firefighter garment with non textile insulation. In: Kuklane, K., Holmer, I. (Eds.), Proceedings of NOKOBETEF 6 and 1st European Conference on Protective Clothing, Stockholm, Sweden. European Society for Protective Clothing, Denmark, pp. 293–295.

Holme, I., 2004. Innovations in performance clothing and microporous film. Tech. Text. Int. 13 (4), 26–30.
Holmer, I., Nilsson, H., 1995. Heated manikin as a tool for evaluating clothing. Ann. Occup. Hyg. 39 (6), 809–818.
Hosseini Ravandi, S.A., Valizadeh, M., 2011. Properties of fibers and fabrics that contribute to human comfort. In: Song, G. (Ed.), Improving Comfort in Clothing. Woodhead Publishing Limited, Cambridge, pp. 61–68.
Hostler, D., Reis, S.E., Bednez, J.C., Kerin, S., Suyama, J., 2010. Comparison of active cooling devices with passive cooling for rehabilitation of firefighters performing exercise in thermal protective clothing: a report from the fireground rehab evaluation (FIRE) trial. Prehosp. Emerg. Care 14 (3), 300–309.
Huang, J.H., 2006. Sweating guarded hot plate test method. Polym. Test. 25 (5), 709–716.
Huck, J., McCullough, E.A., 1988. Firefighter turnout clothing, physiological and subjective evaluation. In: Mansdrof, S.Z., Sagar, R., Nielson, A.P. (Eds.), Performance of Protective Clothing—ASTM STP 989. ASTM International, West Conshohocken, PA, pp. 439–451.
ISO 11092, 2014. Measurement of thermal and water vapor resistance under steady-state condition using sweating guarded hot plate.
ISO 15831, 2004. Clothing—physiological effects—measurement of thermal insulation by means of a thermal manikin.
ISO 7933, 2004. Ergonomics of the thermal environment—analytical determination and interpretation of heat stress using calculation of the predicted heat strain.
ISO 9886, 2004. Ergonomics—evaluation of thermal strain by physiological measurements.
ISO 9920, 2007. Ergonomics of the thermal environment—estimation of thermal insulation and water vapour resistance of a clothing ensemble.
Jin, L., Hong, K.A., Nam, H.D., Yoon, K.J., 2011. Effect of the thermal barrier on the thermal protective performance of firefighter garment. In: Li, Y., Luo, X.N., Liu, Y.F. (Eds.), Proceedings of TBIS2011, Beijing, China. Textile Bioengineering and Informatics Symposium Society, Hong Kong, pp. 1010–1014.
Karaca, E., Kahraman, N., Omeroglu, S., Becerir, B., 2012. Effects of fiber cross sectional shape and weave pattern on thermal comfort properties of polyester woven fabrics. Fibres Text. East. Eur. 20 (3), 67–72.
Katch, V.L., McArdle, W.D., Katch, F.I., 2006. Essentials of Exercise Physiology, fourth ed. Lippincott Williams & Wilkins, Baltimore, MD.
Keiser, C., Becker, C., Rossi, R.M., 2008. Moisture transport and absorption in multilayer protective clothing fabrics. Text. Res. J. 78 (7), 604–613.
Lawson, L.K., Crown, E.M., Ackerman, E.Y., Dale, J.D., 2004. Moisture effect in heat transfer through clothing systems for wildland firefighters. Int. J. Occup. Saf. Ergon. 10 (3), 227–238.
Lee, Y.M., Barker, R.L., 1986. Effect of moisture on the thermal protective performance of heat-resistant fabrics. J. Fire Sci. 4 (5), 315–330.
Lee, J., Kim, E., Yoo, S., Baek, B., Hwang, S., 2007. Development of an intelligent turnout gear for dynamic thermal protection using two-way shape memory alloy. In: 4th International Avantex Symposium for Innovative Apparel Textiles, Germany.
Li, Y., 2001. The science of clothing comfort. Text. Prog. 31 (1/2), 1–135.
Li, Y., Holcombe, B.V., 1998. Mathematical simulation of heat and moisture transfer in a human-clothing-environment system. Text. Res. J. 68 (6), 389–397.
Li, X.H., Lu, Y.H., Li, J., Wang, Y.Y., Zhou, L., 2012. A new approach to evaluate the effect of moisture on heat transfer of thermal protective clothing under flashover. Fibers Polym. 13 (4), 549–554.

Likert, R., 1932. A technique for the measurement of attitudes. Arch. Psychol. 22 (140), 1–55.
Lu, Y., Song, G., Li, J., 2013a. Analyzing performance of protective clothing upon hot liquid exposure using instrumented spray manikin. Ann. Occup. Hyg. 57 (6), 793–804.
Lu, Y., Li, J., Li, X., Song, G., 2013b. The effect of air gaps in moist protective clothing on protection from heat and flame. J. Fire Sci. 31 (2), 99–111.
Mah, T., Song, G., 2010a. Investigation of the contribution of the garment design to thermal protection. Part 2: instrumented female mannequin flash fire evaluation system. Text. Res. J. 80 (14), 1473–1487.
Mah, T., Song, G., 2010b. An investigation of the contribution of garment design to thermal protection. Part I: characterizing air gaps using 3-D body scanning for women's protective clothing. Text. Res. J. 80 (13), 1317–1329.
Mandal, S., Abraham, N., 2010. An overview of sewing threads mechanical properties on seam quality. Pak. Text. J. 59 (1), 40–43.
Mandal, S., Song, G., 2011. Characterization of protective textile material for thermal hazard. In: Fiber Society Spring Conference, Hong Kong.
Mandal, S., Song, G., 2014. An empirical analysis of thermal protective performance of fabrics used in protective clothing. Ann. Occup. Hyg. 58 (8), 1065–1077.
Mandal, S., Song, G., 2015. Thermal sensors for performance evaluation of protective clothing against heat and fire: a review. Text. Res. J. 85 (1), 101–112.
Mandal, S., Song, G., Ackerman, M., Paskaluk, S., Gholamreza, F., 2013. Characterization of textile fabrics under various thermal exposures. Text. Res. J. 83 (10), 1005–1019.
Mandal, S., Lu, Y., Wang, F., Song, G., 2014. Characterization of thermal protective clothing under hot water and pressurized steam exposure. AATCC J. Res. 1 (5), 7–16.
Mandal, S., Song, G., Rossi, R., 2016. Thermal Protective Clothing for Firefighters. Woodhead Publishing.
Manore, M.M., Berry, T.E., Skinner, J.S., Carroll, S.S., 1991. Energy expenditure at rest and during exercise in nonobese female cyclical dieters and in nondieting control subjects. Am. J. Clin. Nutr. 54 (1), 41–46.
Marom, G., Weinberg, A., 1975. Effect of fiber critical length on thermal expansion of composite-materials. J. Mater. Sci. 10 (6), 23–127.
Matsudaira, M., Tan, Y., Kondo, Y., 1993. Effect of fibre cross-sectional shape on mechanical properties and handle. J. Text. Inst. 84 (3), 376–386.
Matusiak, M., 2006. Investigation of the thermal insulation properties of multilayer textiles. Fibres Text. East. Eur. 14 (5), 98–102.
McCarthy, L.K., Marzo, M., 2011. The application of phase change material in fire fighter protective clothing. Fire. Technol 48 (4), 841–864.
McCullough, E., Jones, B., Tamura, T., 1989. A data base for determining the evaporative resistance of clothing. ASHRAE Trans. 95 (2), 316–328.
McCullough, E., Huang, J., Kim, C.S., 2004. An explanation and comparison of sweating hot plate standards. J. ASTM Int. 1 (7), 121–133.
Mecheels, J., Umbach, K.H., 1976. Thermophysiological properties of clothing systems. Melliand Textilber. 57 (12), 1029–1032.
Mecheels, J., Umbach, K.H., 1977. Thermophysiological properties of clothing systems. Melliand Textilber. 58 (11), 76–85.
Meinander, H., 1999. Extraction of data from sweating manikin test. In: 3rd International Meeting on Thermal Manikin Testing, Sweden.
Morozumi, Y., Akaki, K., Tanabe, N., 2012. Heat and moisture transfer in gaps between sweating imitation skin and nonwoven cloth: effect of gap space and alignment of skin and clothing on the moisture transfer. Heat Mass Transf. 48 (7), 1235–1245.

Morris, G.J., 1953. Thermal properties of textile materials. J. Text. Inst. Trans. 44 (10), T449–T476.

Mountain, S.J., Sawka, M.N., Cadarette, B.S., Quigley, M.D., McKay, J.M., 1994. Physiological tolerance to uncompensable heat stress: effects of exercise intensity, protective clothing and climate. J. Appl. Physiol. 77 (1), 216–222.

Murakami, H., Hegemier, G., Maewal, A., 1978. Mixture theory for thermal-diffusion in directional composites with cylindrical fibers of arbitrary cross-section. Int. J. Solids Struct. 14 (9), 723–737.

NFPA 1951, 2013. Standard on protective ensembles for technical rescue incidents.

NFPA 1971, 2007. Standard on protective ensembles for structural firefighting and proximity firefighting.

NFPA 1977, 2011. Standard on protective clothing and equipment for wildland fire fighting.

NFPA 1999, 2013. Standard on protective clothing and ensembles for emergency medical operations.

Ogniewicz, Y., Tien, C.L., 1981. Analysis of condensation in porous insulation. Int. J. Heat Mass Transf. 24 (4), 421–429.

Otsuka, K., Wayman, C.M., 1998. Shape Memory Materials. Cambridge University Press, Cambridge.

Perkins, R.M., 1979. Insulative values of single-layer fabrics for thermal protective clothing. Text. Res. J. 49 (4), 202–205.

Poehlman, E.T., 1992. Energy expenditure and requirements in aging humans. J. Nutr. 122 (11), 2057–2065.

Poehlman, E.T., Horton, E.S., 1989. The impact of food intake and exercise on energy expenditure. Nutr. Rev. 47 (5), 129–137.

Psikuta, A., Wang, L.C., Rossi, R.M.J., 2013. Prediction of the physiological response of humans wearing protective clothing using a thermophysiological human simulator. J. Occup. Environ. Hyg. 10 (4), 222–232.

Ramachandran, T., Manonmani, G., Vigneswaran, C., 2010. Thermal behaviour of ring- and compact-spun yarn single jersey, rib and interlock knitted fabrics. Indian J. Fibre Text. Res. 35 (3), 250–257.

Rao, D.R., Gupta, V.B., 1992. Thermal characteristic of wool fiber. J. Macromol. Sci., Phys. 31 (2), 149–162.

Rees, W.H., 1941. Transmission of heat through textile fabrics. J. Text. Inst. 32 (8), T149–T166.

Rees, W.H., 1946. The protective value of clothing. J. Text. Inst. 37 (7), 132–153.

Rehn, B., 2004. ADAM advanced automotive thermal manikin. http://www.zyn.com/flc/meeting/presentations/Rehn Comfort.pdf (accessed 15.12.14.).

Rengasamy, R.S., Kawabata, S., 2002. Computation of thermal conductivity of fibre from thermal conductivity of twisted yarn. Indian J. Fibre Text. Res. 27 (4), 342–345.

Richards, M.G., Mattle, N.G., 2001. Development of a sweating agile thermal manikin (SAM). In: 4th International Meeting on Thermal Manikins, Switzerland.

Rwei, S.P., Lin, Y.T., Su, Y.Y., 2005. Study of self-crimp polyester fibers. Polym. Eng. Sci. 45 (6), 838–845.

Saris, W.H.M., Vanerpbaart, M.A., Brouns, F., Westerterp, K.R., Tenhoor, F., 1989. Study on food intake and energy expenditure during extreme sustained exercise: the Tour de France. Int. J. Sports Med. 10 (1), 26–31.

Semenniuk, K.M., Dionne, J.P., Makris, A., Bernard, T.E., Ashley, C.D., 2005. Evaluating the physiological performance of a liquid cooling garment used to control heat stress in hazmat protective clothing ensembles. J. ASTM Int. 2 (2), 1–9.

Shalev, I., Barker, R.L., 1983. Analysis of heat transfer characteristics of fabrics in an open flame exposure. Text. Res. J. 53 (8), 475–482.
Shalev, I., Barker, R.L., 1984. A comparison of laboratory methods for evaluating thermal protective performance in convective/radiant exposures. Text. Res. J. 54 (10), 648–654.
Shekar, R.I., Kotresh, T.M., Subbulakshmi, M.S., Vijaylakshmi, S.N., Prasad, A.S.K., 2009. Thermal resistance properties of paratrooper clothing. J. Ind. Text. 39 (2), 123–148.
Shin, Y., Yoo, D.I., Son, K., 2005. Development of thermoregulating textile materials with microencapsulated phase change materials. J. Appl. Polym. Sci. 96 (6), 910–915.
Song, G., 2007. Clothing air gap layers and thermal protective performance in single layer garment. J. Ind. Text. 36 (3), 193–205.
Song, G., 2009. Thermal insulation properties of textiles and clothing. In: Williams, J.T. (Ed.), Textiles for Cold Weather Apparel. Woodhead Publishing, Cambridge, pp. 19–32.
Song, G., Lu, Y., 2013. Structural and proximity firefighting protective clothing: textiles and issues. In: Kilinc, F.S. (Ed.), Handbook of Fire Resistant Textiles. Woodhead Publishing, Cambridge, pp. 520–548.
Song, G., Paskaluk, S., Sati, R., Crown, E.M., Dale, D.J., Ackerman, M., 2011. Thermal protective performance of protective clothing used for low radiant heat protection. Text. Res. J. 81 (3), 311–323.
Speakman, J.B., Chamberlain, N.H., 1930. The thermal conductivity of textile materials and fabrics. J. Text. Inst. 21 (2), T29–T56.
Sun, G., Yoo, H.S., Zhang, X.S., Pan, N., 2000. Radiant protective and transport properties of fabric used by wildland firefighters. Text. Res. J. 70 (7), 567–573.
Tian, M.W., Zhu, S.K., Pan, N., 2012. Measuring the thermophysical properties of porous fibrous materials with a new unsteady-state method. J. Therm. Anal. Calorim. 107 (1), 395–405.
Torvi, D.A., Threlfall, T.G., 2006. Heat transfer model of flame resistant fabrics during cooling after exposure to fire. Fire. Technol 42 (1), 27–48.
Varshney, R.K., Kothari, V.K., Dhamija, S., 2011. A study on thermophysiological comfort properties of fabrics in relation to constituent fiber fineness and cross-sectional shape. J. Text. Inst. 101 (6), 495–505.
Wang, F., Kuklane, K., Gao, C., Holmer, I., 2011. Can the PHS model (ISO7933) predict reasonable thermophysiological responses while wearing protective clothing in hot environments? Physiol. Meas. 32 (2), 239–249.
Wu, Y.S., Fan, J.T., 2009. Measuring the thermal resistance and evaporative resistance using a sweating manikin. Meas. Sci. Technol. 20 (2), 95–108.
Wu, H.J., Fan, J., Qin, X.H., Zhang, G., 2008. Thermal radiative properties of electrospun superfine fibrous PVA. Mater. Lett. 62 (6/7), 828–831.
Wu, H.Y., Zhang, W.Y., Li, J., 2009. Study on improving the thermal-wet comfort of clothing during exercise with an assembly of fabrics. Fibres Text. East. Eur. 17 (4), 46–51.
Wu, Y.S., Fan, J.T., Yu, W.M., 2011. Effect of posture positions on the evaporative resistance and thermal insulation of clothing. Ergonomics 54 (3), 301–313.
Yoo, S., Barker, R.L., 2005. Comfort properties of heat-resistant protective workwear in varying conditions of physical activity and environment. Part I: thermophysical and sensorial properties of fabrics. Text. Res. J. 75 (7), 523–530.

Testing and evaluation of wearable electronic textiles and assessment thereof

4

M. Stoppa, A. Chiolerio
Italian Institute of Technology, Torino, Italy

4.1 Introduction

During the last decade, textiles entered a new era of *smart textiles*. Mainly, smart textiles are defined as textiles that interact with their surroundings. They can be described as textile materials that think for themselves, for example through the incorporation of electronic devices or smart materials (Stoppa and Chiolerio, 2014).

According to functional activity (Sinclair, 2014; Van Langenhove and Hertleer, 2004), smart textiles can be classified in three different categories:

- I generation → able to sense the environment/user, based on sensors (Passive Smart Textiles);
- II generation → reactive, sensing stimuli from the environment, integrating an actuator function and a sensing device (Active Smart Textiles). The actuators act upon the detected signal either directly or from a central control unit;
- III generation →able to sense, react, and adapt their behavior to the given circumstances (Ultra Smart Textiles, UST).

Electronic textiles, known as *e-textiles*, are a subset of smart textiles. They refer to a textile structure with electrical functioning similar to electronics, but physically behaving as a textile at the same time. In general, the development of electronic textiles supports the idea of wearable computing, or electronic devices into garment designs (Van Langenhove, 2007). Generally, e-textiles comprise the following components:

- sensors to detect body/environmental parameters;
- a data processing unit to collect and process the obtained data;
- an actuator that can give a signal to the wearer;
- an energy supply;
- interconnections for both power and signal;
- a communication device that establishes a wireless communication link with a nearby base station.

A variety of physiological and behavioral measurements relevant to clinical rehabilitation have been enabled through the integration of sensors within a fabric substrate (Fleury et al., 2015). For instance, fabric sensors can be used for electrocardiogram (ECG) (Alzaidi et al., 2012), electromyography (EMG) (Benatti et al., 2014), and electroencephalography (EEG) (Löfhede et al., 2012) sensing. In these applications,

Performance Testing of Textiles. http://dx.doi.org/10.1016/B978-0-08-100570-5.00005-0

signal quality has generally been the primary consideration while user comfort and sensor integration have become secondary.

Moving beyond the biosignal sensing, several types of sensors based on textiles can be used for others purposes, such as: temperature sensing (Laukhina et al., 2014), hosting conductive fibers into fabric that responds with a temperature variation; biophotonic sensing (Omenetto et al., 2011) using luminescent elements integrated with normal fabric; sensing movements (Grillet et al., 2008) or strain events using shape-sensitive fabrics; gas detection (Windmiller and Wang, 2013) adopting fibers able to change their electrical properties when a particular chemical agent is presented near the sensor.

Active functionalities could include energy storage (Vatansever et al., 2011), human interface elements (Schneegass et al., 2014), radio frequency (RF) functionality, or assistive technology (Black, 2007). This category also comprises smart textiles with a power generation feature and it can be achieved through piezoelectric elements (Nilsson et al., 2014) that harvest energy from motion, or photovoltaic elements (Lee et al., 2013).

UST are the new ultimate frontier of the innovation in this field. Reached through a successful fusion between textile technology and material science, mechanics, biology, sensor and actuator technology, etc., this innovation is currently ongoing. An example for all could be textiles with embedded stimuli–response polymers, able to change their properties thanks to environmental stimuli, and in the meantime could produce an action, such as drug release. Coupling these smart textiles with electronics and/or computational architectures would lead to fully integrated and autonomous systems (Hu et al., 2012b).

This intelligent yarn will form the basic building blocks of truly flexible, drapeable, and washable textile computers in the near future, creating third-generation wearable computing.

This chapter provides a review of e-textiles, describing the state of the art about research and development of wearable systems reporting the salient characteristics, the manufacturing methods, and the relative functional property characterization and assessment from different levels of integration. Each scientific approach will be followed by some related work carried out by companies, universities, or research institutes.

4.2 e-Textiles manufacturing methodologies and characterization

Fabrics present a hierarchical structure based on fibrous materials. This particular architecture allows the development of e-textiles from different levels of integration and in recent years; several companies and researchers worked to seek the most efficient manufacturing process.

In order to understand the different levels, Fig. 4.1 presents the hierarchical structure of the fabric and the most common processing techniques.

Fibers are the first integration level. They are characterized by having a high length-to-thickness ratio (Needles, 1981). Yarns are the second level of integration,

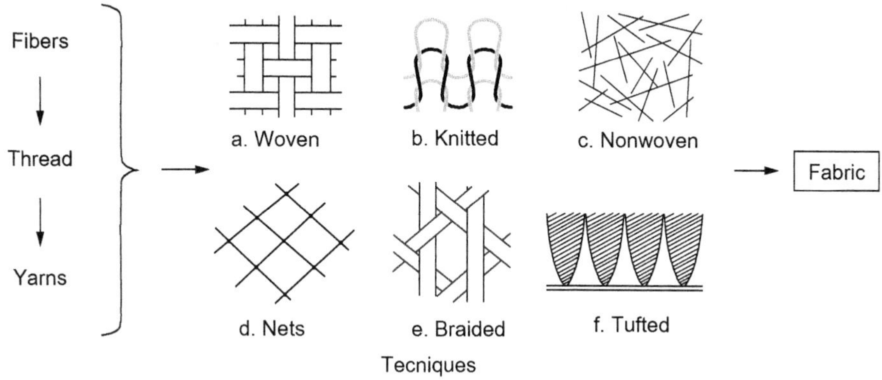

Fig. 4.1 Fabric construction platforms and relative techniques.

and through different manufacturing processes, yarns get turned into fabric, reaching the third level of integration. By coupling different layers of fabric, other levels can be developed (ie, 3D textiles) (Ansar et al., 2011).

Starting from these levels, different e-textile manufacturing methods have been invented and developed. In parallel, different base materials have been tested to achieve the best electrical behavior, near the traditional electronics features, according to the final applications.

4.2.1 Conductive fibers

Conductive fibers are a mix of electric wires and the textiles world, with attributes of each. These fibers consist of a nonconductive or less-conductive substrate, which is then either coated or embedded with electrically conductive elements (Kallmayer and Simon, 2012). Straddling the worlds of textiles and wires, conductive fibers are sold either by weight or length, and their geometry/density is measured in denier and AWG (Winterhalter et al., 2005).

Initially, conductive fibers were mainly used in technical areas such as: clean room garments, military apparel, medical application, and electronics manufacturing (Resistat, 2015). They can have a variety of functions, like antistatic applications (Sophitex, 2015), ElectroMagnetic Interference shielding (LessEMF, 2015), electronic applications, infrared absorption or protective clothing in explosive areas (Mcfarland et al., 1999).

In order to use these conductive fibers in the preceding applications, it is important that they should be strong, flexible, environmentally stable, and resistant to chemical agents.

Especially for clothing, tactile properties such as stretch, shear, and handle are quite relevant. The tactile comfort is related to mechanical interaction between the clothing material and the human body and is an intrinsic and essential performance requirement in clothing. Fabric handle is concerned with the feeling of the material and so depends upon the sense of touch for a subjective evaluation of an expert (Sülar and Okur, 2008).

For this reason the fibers should be fine and fabrics should have a low weight per unit area (not more than 300 g/m^2) (Statex, 2015). These demands are inconsistent with the materials and geometries that are needed for electrical conductivity purposes. Fig. 4.2 shows different types of conductive (black) and insulating (white) structures of conductive fibers. Each of those could have different applications, for example a shell conductive fiber is best applied where there is no need for bulk signal transmission (low frequency and high power, while rising the frequency for the so-called skin effect the conduction is sustained in the very superficial layer of a conductor), a wholly conductive fiber is not supposed to suffer from surface modification/oxidation, a core conductive fiber is normally the solution for avoiding surface modification/oxidation, and so on.

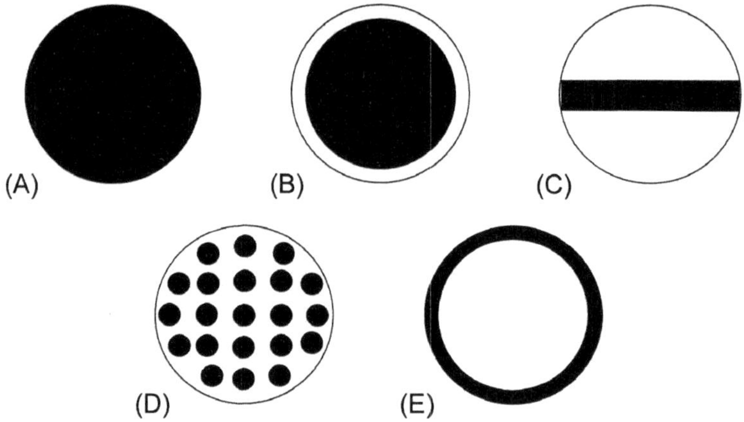

Fig. 4.2 (A) Wholly conductive fiber; (B) core conductive fiber; (C) sandwich-type fiber; (D) coated yarns, and (E) shell conductive fiber.

Different methods of conductive fiber manufacturing are available today, such as melt spinning, wet spinning (Pereira et al., 2000), electrochemical processing (Araki et al., 2011), or coating (Yamashita et al., 2012) of conventional insulating materials with conductive materials.

Following the manufacturing processes adopted, the conductive fibers can be divided into *intrinsic* or *extrinsic* (Bashir, 2013). These processes show different advantages and disadvantages as summarized in Table 4.1. The *intrinsic* conductive fibers are made by materials featuring high electrical conductivity, and in particular:

- *metallic* and *metallic alloy fibers* are very thin metal filaments with diameters ranging from 1 to 80 μm and they are made by the bundle-drawing or shaving process (Meoli and May-Plumlee, 2002). Metal content for clothing is difficult to process and also reduces the comfort of the final product. Their inertness degree, in other words, their likeliness to be sensitive to washing or sweating, depends on the metal used to create the filament. They cannot provide uniform heating and their brittle characteristics can damage spinning machinery over time (Araki et al., 2012);
- *carbon fibers* have a graphite-like structure with a conductivity comparable to that of metal (10^4–10^6 S/cm) (Dalmas et al., 2006);

Table 4.1 **Advantages and disadvantages using intrinsic and extrinsic process with different base materials**

Process manufacturing	Material	Advantages	Disadvantages
Intrinsic	Metallic and metallic alloy fiber	High conductivity (10^6 S/cm) and mechanical resistance[a]	Low flexibility, stiffness, high weight, and poor weaving properties[b]
	Carbon fibers	High conductivity, strength, stiffness, low weight, and high fatigue resistance[c]	Hard integration into knitted and weaved structure and harmful to health[c]
	Instrinsically conducting polymers (ICPs)	Medium conductivity (~200 S/cm) and low weight[d]	Poor mechanical strength, brittleness, and difficult processing[e]
Extrinsic	Conductive filled fibers	High conductivity[f]	High manufacturing cost[g]
	Conductive coated fibers	High conductivity (10^6 S/cm) and mechanical properties[h]	High manufacturing cost, stiffness, brittleness, and high weight[i]

References:
[a]Tibtech (2015).
[b]Meoli and May-Plumlee (2002).
[c]Hunt et al. (2012).
[d]Baik et al. (2009).
[e]Pomfret et al. (2000).
[f]Bigg Battelle (1977).
[g]Shirakawa et al. (1977).
[h]Xue et al. (2004).
[i]Sen (2007).

- *Intrinsically conducting polymers (ICPs)* are organic conductive materials based on polyaniline (PANI), poly(3,4-ethylenedioxithiophene)polystyrene sulfonate (PEDOT:PSS), or polypyrrole (PPy) (Ding et al., 2010; Bocchini et al., 2013).

Taking into account the *extrinsic* conductive fibers, they are made by the combination between conductive and nonconductive materials. Special treatments involve the mixing, blending, or coating process. These approaches guarantee good electrical and mechanical properties of the treated materials. These fibers can be divided into:

- *conductive filled fibers*, which are made by adding conductive fillers (eg, metallic powder, metallic nanowires, carbon nanotubes (CNTs) or ICPs) into nonconductive polymers such as polypropylene, polystyrene, or polyethylene (Thongruang et al., 2002). Melt spinning and solution spinning are the common processes to develop this kind of fiber. The solution spinning process guarantees fibers with improved electrical and mechanical properties over those produced by the melt spinning;

- *conductive coated fibers* are produced by coating insulating materials with conductive ones (eg, carbon black, metals, CNTs or ICPs) (Xue et al., 2004). The properties of these fibers depend on the type of conductive materials used and the relative manufacturing method. Metallic coatings, sputtering, vacuum deposition, and filling or loading fibers are the most common methods applied. Adopting carbon black, CNTs and ICPs as a thin layer to a fiber surface through solution casting, inkjet printing, chemical vapor deposition or vapor phase polymerization methods (Im and Gleason, 2007) improve the quality of the conductive fibers.

Electrical and mechanical properties are correlated to each other and the resistance variation of spun conductive fibers, depending on tensions applied. Flexibility is defined as the resistance to permanent deformation under stresses, such as folding or bending. It can be improved through textile processes such as spinning or twisting because the overall geometry of the yarn is a prior factor to those of individual fibers. Generally, the ASTM D1388 (or Federal Test Method Standard 191A-5200 for USA) hanging loop test has been the most well-known test method to evaluate bending behavior of the fabrics (Materials, 2012). A typical machine is the Taber Fabric Stiffness Tester that provides a quick and accurate test method for flexural rigidity employing the principle of cantilever bending (Taber Industries, 2015). This method has been properly adapted for textiles that host metal fibers and metalized polyester shielding, due to the fact that the materials increase their stiffness in comparison to any conventional textile. The reference parameter, bus stiffness (K_b (N/in.)), is calculated as the ratio between the force (F) applied and the relative displacement Δl obtained (see Eq. 5.1).

$$K_b = \frac{F}{\Delta l} \tag{5.1}$$

Another parameter to evaluate the bending properties is the sewability of multiple conductive fibers. The method adopted for this evaluation is the curl test described previously (no International Standard available); it is able to check the residual curling of a conductive wire compared to a conventional sewing thread (Orth, 2002). As mentioned previously, adding conductive properties with a metal part, the mechanical properties are affected and the curl test allows us to understand if a conductive wire is eligible for machine sewing or not.

Washability is another relevant feature to consider and it is related to chemical resistance against moisture and detergents as well as physical resistance against mechanical stresses and high temperature. Generally, washing tests are conducted in order to accelerate aging for conventional electronics (Zeagler et al., 2013). The test comprises the use of a standard upright agitator machine (eg, GE Spacemaker Model) and a standard detergent, in order to simulate an ordinary washing. Currently, many conductive fibers have been tested with this method and the results showed that the electrical resistance may increase by three orders of magnitude after the wash test (eg, from 3 kΩ to 3 MΩ for 30 cm of conductive strip made with Shieldex fibers size 40 22/7 PET).

Nowadays, several companies produce conductive fibers adopting different manufacturing methods. Textiles can be simply intertwined with filaments of stainless steel or other metals, or they can be covered with metals such as aluminum (Al) and magnesium (Mg). Often in literature, no distinction is made between metal wires and metal fibers. However, the Sprint Metal Company (Sprint Metal, 2015) discerns metal fibers and wires according to their diameter. While a fine wire has a diameter between 30 μm and 1.4 mm, a metal fiber possesses a diameter of 2–40 μm (Ugitech, 2015).

CanShielding company develops lightweight aprons based on carbon (Fujimori et al., 2011) and silver fibers (Dordevic, 1992), to protect X-ray technicians with improved comfort and wearability (Canshielding, 2015). This kind of fiber allows production of fabric that partially absorbs radar signals.

Swiss company Elektrisola Feindraht AG (Elektrisola, 2015) fabricates metal monofilaments that can be blended with all sorts of fibers or that can be directly used in weaving and knitting. Importantly, the material has different electrical properties (see Stoppa and Chiolerio, 2014). The products range from copper (Cu) and silver-plated copper (Cu/Ag) filaments, to brass (Ms) and silver-plated brass (Ms/Ag) filaments, to aluminum (Al) filaments, to copper-clad aluminum filaments (Elektrisola, 2015).

The company Swiss-Shield® is specialized in producing metal monofilaments which are incorporated into base yarns like cotton, polyester, polyamides, and aramides. Fig. 4.3A shows a typical conductive yarn with base-fibers and a metal monofilament twisted around them (Swiss Shield, 2015).

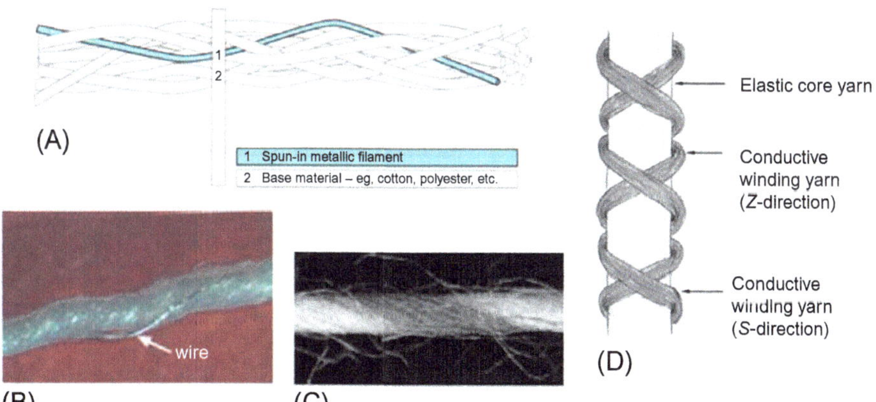

Fig. 4.3 (A) Schematic of conductive metal fiber twisted with the natural or synthetic fibers. From Swiss Shield. (B) Twisted metal wire: The metal wire is twisted around the polymer yarn. (C) Metal fibers: The conductive yarn consists of metal multifilament. (B) and (C) from Locher, I., 2006. Technologies for System-on-Textile Integration. Available at: http://e-collection.library.ethz.ch/view/eth:28457 (accessed 20.01.14). (D) Conductive winding yarns wound around an elastic core yarn in *S*- and *Z*-direction. From Schwarz, A., Kazani, I., et al., 2011. Electro-conductive and elastic hybrid yarns—the effects of stretching, cyclic straining and washing on their electro-conductive properties. Mater. Des. 32 (8), 4247–4256.

4.2.2 Conductive fabrics

After understanding the electrical properties and the manufacturing methods of the conductive fibers, to achieve a conductive fabric, the next level of integration is the conductive yarn.

Conductive yarns can be either based on nonmetallic conductive materials, such as carbon fibers, or metallic materials, as metal fibers. Such yarns can include a set of metal filaments, for example, stainless steel filaments, which can be twisted around each other. By twisting a fine metal wire around a multifilament fiber, the electrical conductivity of the yarn is achieved with an improvement of the mechanical stability (Fig. 4.3B and C).

The spiral-shape of the metal wire allows a flexibility of the whole yarn and in particular, when the yarn is stretched, the metal wire is able to follow the deformation. Using the conductive thread approach, no additional step after manufacturing of the fabric is required to establish conductivity. The conductivity of these threads lies in the range of 10–500 Ω/m (Locher, 2006).

In order to evaluate the resistance shift during a stretching test, generally the measurements are performed with a strain tester, such as Zwick/Z010. The sample yarn is linked and simultaneously connected with a digital multimeter for an analog measurement, or with a microcontroller/read-out circuit in the case of a digital one (Schwarz et al., 2011b). Oh et al. (2003) adopted the conductivity (S/cm) variation as an electrical parameter in relation to elongation. The changes of mechanical properties of the conductive yarns were investigated with a tensile testing machine by a standard test method (ASTM D 751-95) coupled with a digital multimeter. Other evaluations with this experimental setup are focused on aging and hysteresis properties of the material (Guo and Berglin, 2009).

A mixture of staple fiber materials and a certain amount of electrically conductive staple fibers can be mixed and spun into the yarn to achieve the electrical conductivity required. Depending on the metal content, these yarns have, more or less, textile and metallic properties at the same time. Adding only a small percentage of electrically conductive fibers, the result is antistatic clothing, featuring low conductive properties (McCann, 2013).

Blum (2000) describes the manufacturing of a composite yarn of textile roving and monofilament metal thread. During the yarn-spinning process on a ring spinner, a coated metal wire is added centrally to the roving. In the thermal treatment following the spinning process, the melting coating adheres the central wire to the spun textile sheathing. The main issue with this process is the stiffness of the yarns. Moreover, during the manufacturing process of the conductive fabric, a precise position of the conductive yarns within the fabric is not predictable. The helical path around the yarn involves this issue. In terms of electric conductance, this method allows production of fabric features with about 17.8 Ω/m (Cottet et al., 2003).

Another method of achieving conductivity is coating yarns with a metal layer. Adopting an electrochemical process, a fine metal layer such as copper (Cu), silver (Ag), or gold (Au) can be deposited on the surface of the fibers (Schwarz et al., 2009). The advantage of this technique is the low electrical resistance about 50 Ω/m. On the other hand, usually the disadvantage regards the mechanical propriety, because the metallic coating is sensitive to breakage during the deformation of the yarn or during the textile production process. To solve this issue Schwarz et al. (2011b) applied the hollow spindle technology to produce electroconductive and elastic yarns, with enhanced elastic and stable electrical–mechanical properties after exposure to cyclic straining and washing. Fig. 4.3D shows the design of the conductive yarn produced by the hollow spindle process used for washing and stretching tests.

The electrical resistance remains constant (<4 Ω/m for conductive yarns based on copper) over elongation levels up to 100%, and after 25 washing cycles, only the silver-based yarns do not change their electrical resistance. The yarns with the same structure, but based on copper and stainless steel, show an increase in their resistance after the washing test (Schwarz et al., 2011a). The stainless steel and copper yarns were evaluated by quantifying their electrical resistances stability before and after washing with four-point probe techniques. This method is typically used to measure electrical resistance of a thin film or diffusion layer on an insulating material (Schroder, 2006) and the experiments have been carried out with the following setup:

- voltmeter (Keithley 195A Digital Multimeter with resolution of 0.001 mV);
- amperometer (Solartron Schlumberger 7150 Plus Digital Multimeter with resolution of 0.1 mA);
- DC current source (RS PL-series).

The washing tests followed the International Standard ISO 6330:2000 (Washing, 2000) procedure with a Wascator FOM71 CLS washing machine.

As an intermediate approach, introduced by the development of nanotechnologies and their application to the e-textile field, there is the coating of fibers (cotton, polyester, and nylon) with silver nanowires (Ag NWs), in other words, huge aspect ratio monocrystalline silver filaments, typically 15 μm in length and 35 nm in diameter. By properly coating the textile fibers by means of a net of NWs, and annealing them at 150°C for 15 min, one obtains a partial sinterization of the network and the formation of conduction bridges across the NWs. This allows the realization of a conformal, partially transparent, though and flexible coating that is able to convey electrical signals/energy, reaching 8 Ω/m able to survive to 200 bend cycles and to have unchanged properties after 5 washing cycles (see Fig. 4.4) (Atwa et al., 2015).

Moving to the next integration level, we consider that weaving, stitching, and embroidery of conducting yarns in or on the fabric are currently used to achieve electrical properties. In principle, standard manufacturing methods are used to incorporate the conducting yarn in the textile, such as:

- manually attaching conventional wires or sewing conductive thread (Orth, 2002), the yarn is added by guiding it with a needle through the textile;
- machine embroidering of conductive thread (Post et al., 2000), in which the yarn is fixed on the surface of the textile by a separate thread;
- replacing nonconductive fibers with conductive ones (Nakad, 2003).

In these methods, the conductive yarn is thus applied after the textile has been made, or even after the product has received its actual shape. Instead, with a weaving on a loom, the conducting yarn is woven simultaneously with the entire textile.

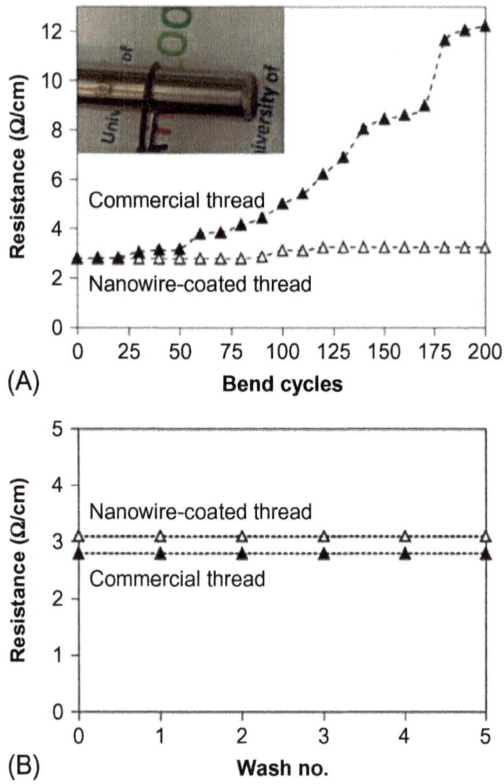

Fig. 4.4 Resistance of a NW-coated nylon thread and commercially available conductive thread after cycling of repeated bends over a 6 mm radius of curvature (A) and of repeated washings (B).
From Atwa, Y., Maheshwari, N., Goldthorpe, I.A., 2015. Silver nanowire coated threads for electrically conductive textiles. J. Mater. Chem. C3, 3908

In principle, the textiles are meant to be worn, washed, stored, and folded, thus the improvement of the electronic textiles shall not affect the original properties. In particular, breaking points between electric components and conductive yarn, or changes of the electrical property of the conductive yarn after mechanical stress, are taken into account. Fig. 4.5 depicts the electrical and mechanical response of a single conducting yarn, compared with a single fabric sample (nonconducting multifilament threads based on polyester) as a function of the strain applied, where de Vries and Peerlings (2014) showed that mechanical failure precedes electrical failure. Moreover, the researchers observed a change of the strains' characteristic due to the weaving pattern.

The British company Baltex uses the knitting technology to incorporate metal wires in textile structures. Their fabrics, which they commercialize under the name Feratec®, can be used mainly for two purposes, namely "heatable" textiles and electromagnetic shielding materials (Baltex, 2014).

The Danish company Ohmatex develops weaving electronics into fabrics, electronic conductors in clothing, operating panels in textiles (soft keyboards, displays,

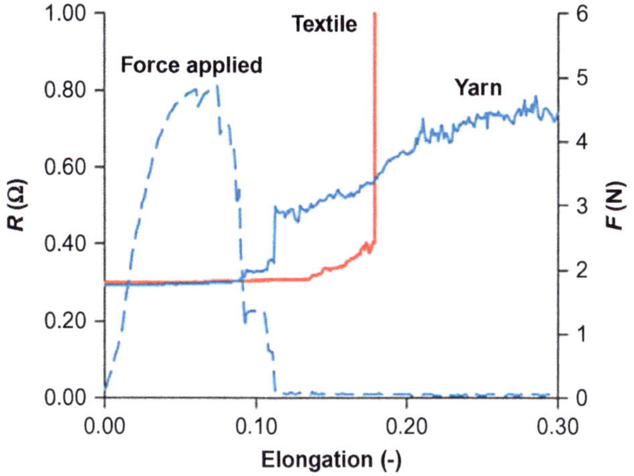

Fig. 4.5 Comparison between the resistance of conductive yarn and a textile sample versus elongation. Blue dashed line shows the force applied to both wires during the elongation phase. From De Vries, H., Peerlings, R., 2014. Predicting conducting yarn failure in woven electronic textiles. Microelectron. Reliab. 54 (12), 2956–2960.

etc.) and microsensors. The conductive yarn that they use is a copper thread, plated with a silver layer and coated with polyester (Ohmatex, 2015).

4.2.3 Conductive inks

Conductive coating is an extrinsic alteration to the fabric substrates and it can be achieved by conductive inks.

It should be specified that the conductive inks must contain an appropriate highly conductive metal precursor such as Ag, Cu, or Au nanoparticles (NPs), alloys of the cited metals, core/shell systems, and a carrier vehicle. Most of them are water based. These specialized inks can be printed onto various materials, among them textiles, to create electrically active patterns (Stoppa and Chiolerio, 2014).

In term of manufacturing process, sheet-based inkjet and screen printing are the best for low-volume, high-precision work and they allow printing of conductive material on different substrates, such as natural or synthetic fabric.

The inkjet technique is flexible and versatile, and can be set up with relatively low effort (Parashkov et al., 2005). It requires minimal material consumption, and it's not necessary to wear a mask. The technique is an additive process without chemical etching and pollutant agents.

However, to obtain a high performance conductive strip via inkjet printing, some technical aspects need to be considered. The conducting layers can be very thin, about 1 μm, requiring measurement with a scanning electron microscope (SEM) and not with a simple rugosimeter or profilometer, and hence, the rough, uneven, and porous surface of the fabric can be an issue (Merilampi et al., 2010). The ink is deposited via droplets and the viscosity is a significant parameter to be considered, because organic or inorganic inks, with dielectric or metal particles dispersed, can clog the nozzle. To overcome this issue, adopting a multiple nozzle system and a prestructured substrate, the resolution and productivity are improved (Blayo and Pineaux, 2005). Also to limit clogging, the dispersed particles should be smaller than one-tenth of the nozzle diameter, as a rule of thumb.

The conductive inks allow a uniform and flat conductive coating on the textile, however, the coating can affect the mechanical properties of the fabrics in terms of tensile, shear, and bending properties (Farboodmanesh et al., 2005) due to the change in yarn mobility after the coating is applied. These changes depend on the viscosity, uniformity, porosity, and flexural rigidity of the conductive material (Dubrovski and Cebasek, 2005). Inks may also contain additives which are used to tune or to add specific ink properties (Tiberto et al., 2013), thus increasing performance in relation to the final application (Perelshtein et al., 2008).

In comparison to a standard substrate for electrical purposes (eg, FR4, polyimide, etc.), a fabric has a roughness surface that can affect the electrical properties of the electrical coating made by inkjet printing. Chauraya et al. (2013) developed a wearable antenna using inkjet printing with the aim of overcoming the problem related to the nonuniform textile surface. They printed two silver layers with a thickness of 3 μm on the stretchable fabric, with a subsequent curing process. One of the two conductive silver layers was chosen as an "interface layer" with the polyester cotton fabric, and the other one was used to improve the conductivity and the pattern definition. Fig. 4.6 shows a cross-section of the yarn, illustrating the layer separation. A thin and uniform intermediate layer allows the yarn to overcome the nonuniform textile surface issue. This layer turns out to be the new surface of the fabric.

Generally, using the inkjet printing techniques, the inks should respect the following requirements (Stassi et al., 2014):

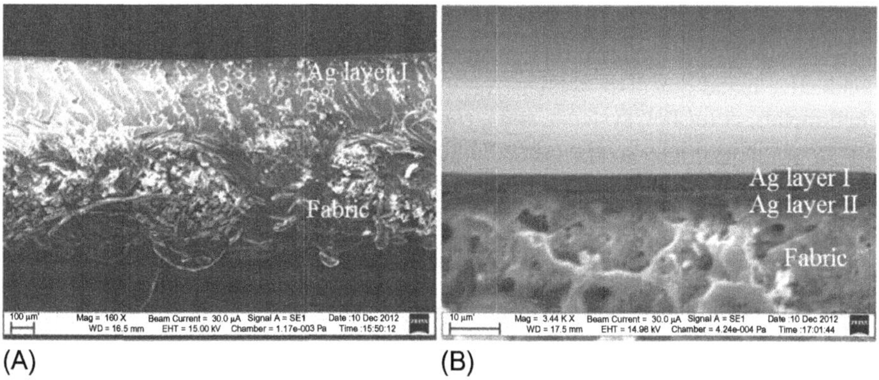

Fig. 4.6 SEM images of printed layers: (A) the interface layer and (B) the inkjet printed layer. From Chauraya, A., et al., 2013. Inkjet printed dipole antennas on textiles for wearable communications. Microw. Antennas Propag. 7 (9), 760–767.

- high electrical conductivity;
- resistance to oxidation;
- dry out without clogging the nozzle during printing;
- good adhesion to the substrate;
- lower particle aggregation;
- suitable viscosity and surface tension.

After inkjet printing of a metal NP-based ink, in order to form a conductive printed pattern, particles must be sintered to create continuous connectivity between them and obtain electrical percolation (for a detailed discussion on sinterization see Chiolerio et al., 2012). Sintering is the process of welding particles together at temperatures below the corresponding bulk metal melting point, involving surface diffusion phenomena rather than phase change between the solid and the liquid. For instance, with inks based on gold NPs (~1.5 nm diameter), the melting temperature was experimentally found to be as low as 380°C, while for inks based on silver NPs (15–20 nm in diameter), a complete sintering can be obtained down to 180°C (Camarchia et al., 2014).

Screen printing is another appropriate technique to develop e-textiles using conducting inks. As shown in Fig. 4.7A, the screen printing procedure, based on a stencil process, comprises the printing of a viscous paste through a patterned fabric screen and is usually followed by a drying process. This allows the ink to develop a pattern and thick layers from pastelike materials. Screen printing also makes integration with planar electronics simpler than with conductive yarn systems. This method can be applied to flat or cylindrical substrates with a resolution of the conductive line lower than 100 μm. By optimizing process, conditions, and materials, the resolution may decrease to 30 μm line/space on thin, flexible substrates (Numakura, 2008). Moreover, screen printing can use inks based on organic semiconductors, and can be printed with a curing temperature of 100°C for PEDOT:PSS (Ionescu et al., 2012).

Although inkjet and screen printing are versatile, they do have their limits. During the printing process, one limit of the silver paste is the thickness. To achieve high

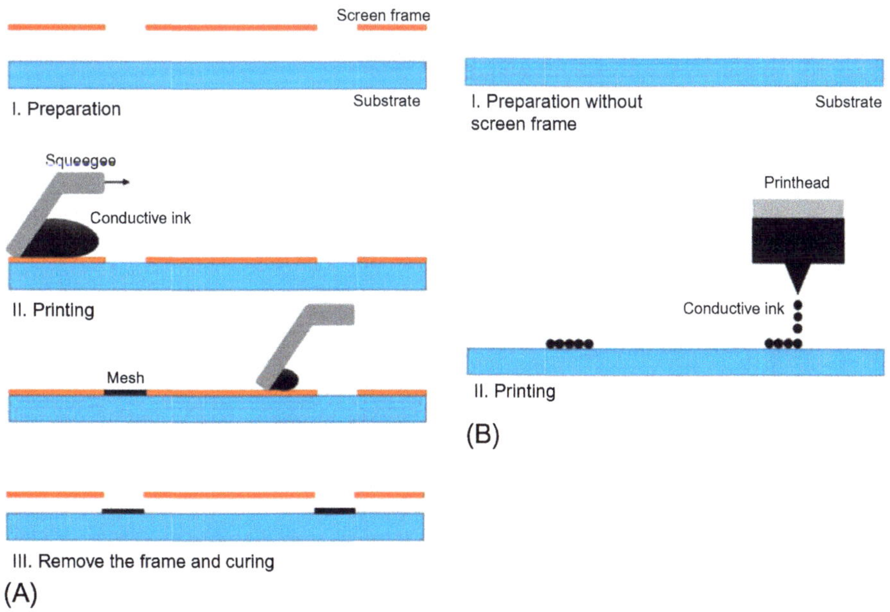

Fig. 4.7 Comparison between (A) screen printing and (B) inkjet printing process.

conductivity and uniformity of the printed line, it is necessary to apply several passes (overprints), decreasing the electrical resistivity as shown in Fig. 4.8. After five printing passes the result is better ink penetration of the fabric. The measurements have been conducted for each printing pass by using time domain reflectometry (TDR) measurements.

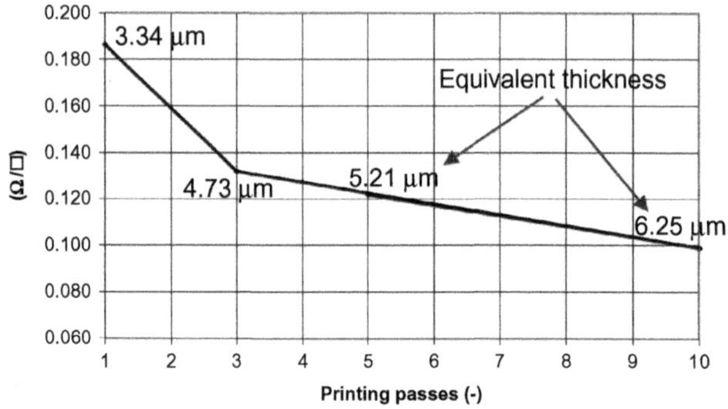

Fig. 4.8 Electrical resistivity versus printing passes highlighting the thickness obtained.
From Locher, I., 2006. Technologies for System-on-Textile Integration. Available at: http://e-collection.library.ethz.ch/view/eth:28457 (accessed 20.01.14).

In addition, several passes affect the mechanical properties, and risk interrupting the electrical conductivity with creasing of the fabric. Fig. 4.9 shows the conductivity changes for the number of creasing interactions with two different thickness levels (Locher, 2006).

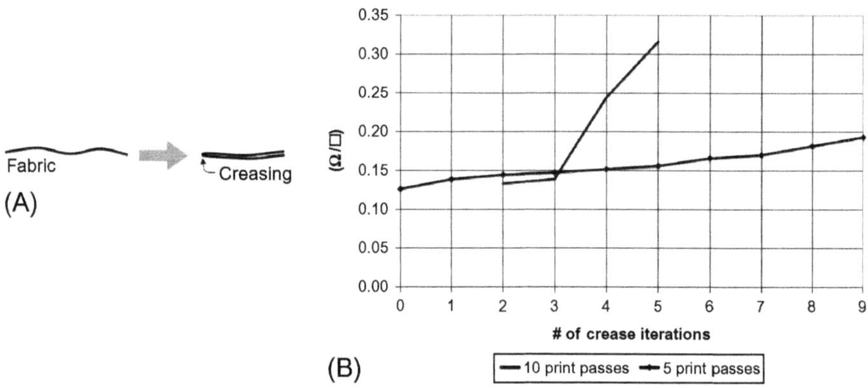

Fig. 4.9 (A) Creasing of the fabric with printed transmission lines; (B) DC resistance changes versus number of creasing iterations.
From Locher, I., 2006. Technologies for System-on-Textile Integration. Available at: http://e-collection.library.ethz.ch/view/eth:28457 (accessed 20.01.14).

A group of researchers of Istituto Italiano di Tecnologia—Center for Space Human Robotics, Politecnico di Torino—Applied Science and Technology Department, in collaboration with a spin-off company, Politronica Inkjet Printing, developed a high-density surface EMG (HD-sEMG) sensor matrix by inkjet printing with a silver NP-based ink on a flexible polyimide substrate. In terms of electrode-skin contact impedance, the results obtained are comparable with commercial electrodes (Scalisi et al., 2015). The HD-sEMG matrix in Fig. 4.10 shows the flexibility of the manufactured electrode matrix (8 × 8), with a diameter of a single electrode of about 8 mm; the high-surface area is achieved using a nanocomposite ink. The results show an intrinsic impedance of the newly printed electrodes of roughly 20 Ω, compared with commercial electrodes that resulted in about 2 Ω of impedance. The tests have been carried out with the Potentiostat 700D CH instruments in order to measure the impedance of the electrodes in their dry conditions on eight different subjects. The impedance of the electrodes in their usage was tested between 10 and 1000 Hz.

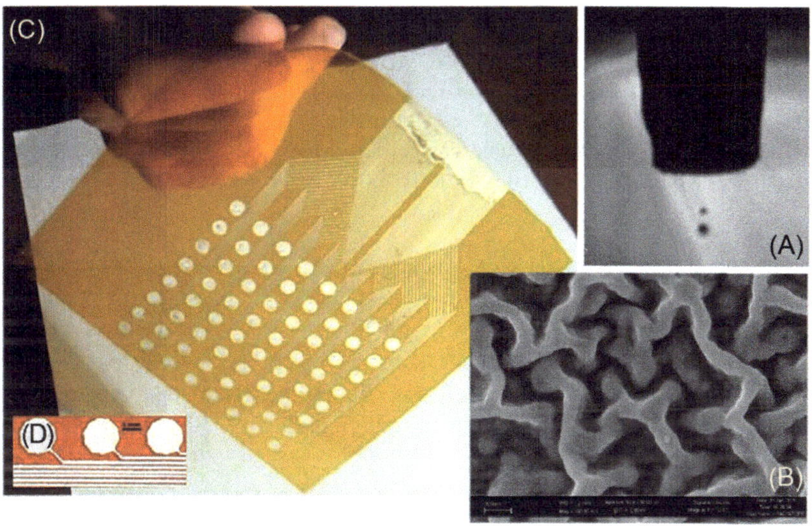

Fig. 4.10 Inkjet printing of the pattern: (A) Ag NP ink droplet ejected by the nozzle. (B) Surface of the nanocomposite ink after solvent removal, showing a high specific area. (C) Flexible electrode 8 × 8 matrix printed on polyimide foil after thermal and insulation treatments. (D) Optical micrograph of some Ag lines (width ~0.5 mm) and electrodes (diameter 8 mm).
From Scalisi, R.G., et al., 2015. Inkjet printed flexible electrodes for surface electromyography. Org. Electron. 18, 89–94.

Remaining in the physiological monitoring field, the National Textile Center at North Carolina State University (Raleigh, NC, USA) developed a garment able to measure ECG, heart rate, respiration, and temperature. They adopted conductive ink to develop the sensors and the conductive line inside a common garment. The electric circuit was printed on nonwoven fabric and different conductive inks were tested, such as Evolon® by Freudenberg KG, Tyvek® by DuPont™, FiberWeb Resolution™ Print Media by BBA FiberWeb™, Precisia LLC and Creative Materials Inc. (Karaguzel et al., 2009), during washing trials. Fig. 4.11 depicts the impedance values of three

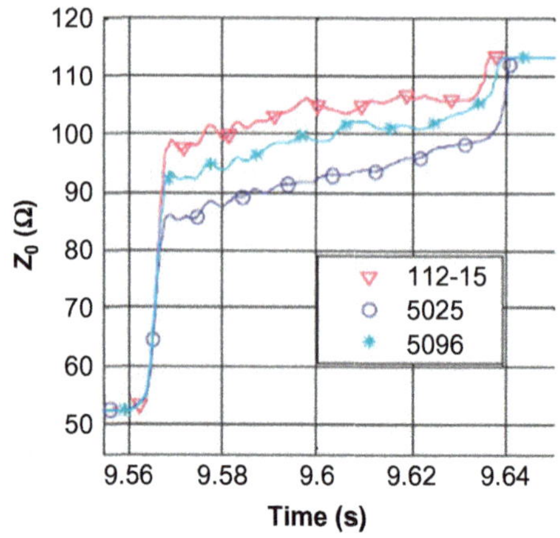

Fig. 4.11 Three different conductive inks (CMI 112-15, DuPont 5025, DuPont 5096), with a similar sheet resistivity (~0.01 Ω/sq./mil), show different impedance value after five washing cycles.
From Karaguzel, B., et al., 2009. Flexible, durable printed electrical circuits. J. Text. Inst. 100 (1), 1–9. Available at: http://dx.doi.org/10.1080/00405000802390147 (accessed 09.01.14).

different conductive inks after five washing cycles. The electrical properties of the printed lines were tested before and after washing, determining the characteristic impedance of the 10 cm long coplanar waveguide lines (CPW) using a Tektronix 11801 B Digital Sampling Oscilloscope in TDR mode. The probe was placed on the CPW while the other end of the line was left open during measurements. DC resistance of the transmission lines was measured using a digital Fluke 77/BN Multimeter.

4.2.4 Planar fabric circuit board

Several research results report the integration of electronics with textiles, however the next step of integration regards the IC components with fabric. For instance, Jung et al. (2003) and Post et al. (2000) placed a silicon chip on a flexible plastic board on a fabric, which resulted in a stiff feeling in the clothes due to the flexible silicon board (eg, Kapton®), in addition to the long integration process and the durability.

Planar thin-film technology is applied directly on the fabric to develop a FCB. It allows the implementation of a circuit board on a plain fabric patch for wearable electronic applications. It features a soft and flexible impression, just as normal clothes do.

Silk screening of conductive epoxy or gold (Au) sputtering technique allows the deposition of planar conductive lines directly on the fabric patch. First, the circuit board is silkscreen printed on the fabric, then the IC is wire-bonded to the patterned electrodes or other components. Finally, the IC is molded with nonconductive epoxy, improving the mechanical properties in the welding points (Kim et al., 2008). For the integrity of the package with the fabrics, the tensile strength test is performed by straining the planar fabric circuit board (P-FCB) along the *x*-direction, *y*-direction, and the diagonal direction. The maximum load is 85.1 N/cm in the x-direction, 85.2 N/cm in the y-direction and 99.3 N/cm in the diagonal direction, otherwise the result is a breaking of the bonding or welding points.

The measured sheet resistance of the silk screen pattern on the fabric (cotton) ranges between 0.01 and 60 mΩ/m, as measured by a standard four-point probe. This value remains unchanged after 50 washing cycles (Kim et al., 2010). The resistance measurement yielded results of the proposed via and conductive adhesive and the frequency response of a P-FCB transmission line (15 cm long and 1 mm wide). Its bandwidth is 80 MHz, and this is sufficient to deal with biosignal processing. The analyses were performed with a vector network analyzer (Lee et al., 2010). To get an average value, as many as 100 connections were formed and measured. The resistances of the proposed via and conductive adhesive measured 0.24 and 0.34 W, respectively (Kim et al., 2010).

Researchers at the Electronics Department and the Wearable Computing Laboratory at the ETH in Zürich demonstrated another approach to develop a fabric circuit board. They started from a plain woven textile structure consisting of polyester (PET) monofilament yarn twisted with copper thread and they developed a hybrid fabric called PETEX (Locher, 2006). In detail, PET yarns have a diameter of 42 μm and copper alloy wires have ~50 μm (Fig. 4.12A).

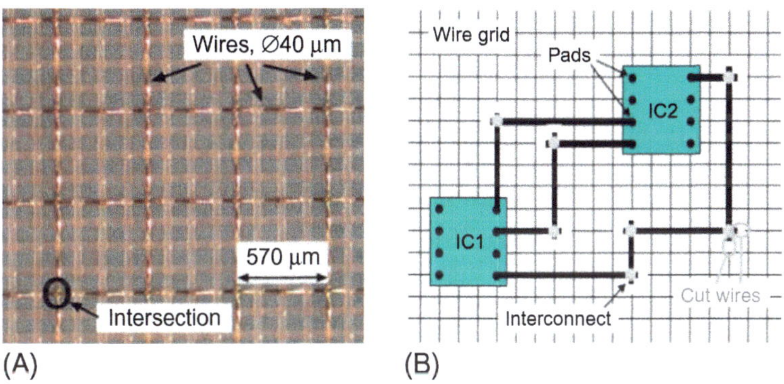

Fig. 4.12 (A) PETEX and (B) method to integrate circuits in a PETEX structure.
From Locher, I., 2006. Technologies for System-on-Textile Integration. Available at: http://e-collection.library.ethz.ch/view/eth:28457 (accessed 20.01.14).

Developing an ad hoc structure, they were able to design a custom textile circuit, as shown in Fig. 4.12B. Generally, the fabric has a regular structure (warp and weft), however in microscale it is not possible to obtain a geometric repeatability of the position of the contact point between the conductive fibers, as shown in Fig. 4.13. This issue, appreciable with a common optical microscope tool, becomes critical when an electrical component needs to be soldered on the e-textile.

4.2.5 Textile sensors

In recent decades a significant breakthrough has been achieved in exploring and enhancing capabilities of textiles in response to environmental stimuli (Castano and Flatau, 2014).

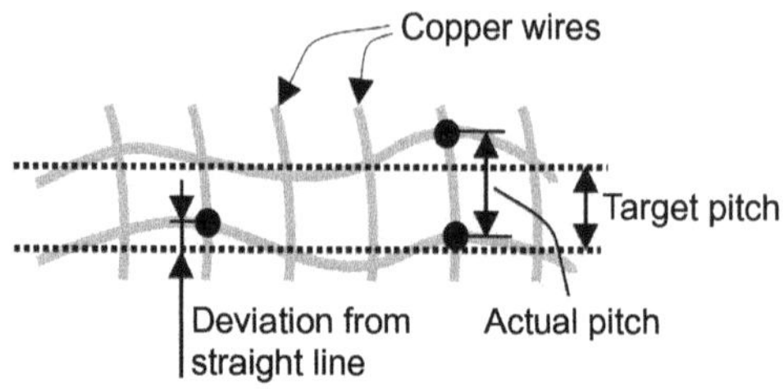

Fig. 4.13 Misalignment of the contact points with the target pitch.
From Locher, I., 2006. Technologies for System-on-Textile Integration. Available at: http://e-collection.library.ethz.ch/view/eth:28457 (accessed 20.01.14).

Conductive textiles that change their electrical properties as a result of the environmental impact can be used as sensors. Smart textiles possess the properties of conventional textile materials and carry additive functional values. Typical examples are textiles that react to deformations such as pressure sensors, stretch sensors, and breathing sensors. Different physical principles are adopted to reach the same purpose, such as capacitive or resistive behavior of the textile sensor. On the other hand, biochemical, optical, temperature, humidity, and biopotential sensors can be made with smart textiles.

4.2.5.1 Pressure sensors

The localized detection or distributed pressure, integrated into textiles and clothes, has been exploited in several application fields (Ashruf, 2002). In healthcare and sports, pressure sensors embedded in shoes enable the analysis of gait. Piezoresistive sensors have also been applied to car safety belts for monitoring cardiorespiratory patterns in subjects (Hamdani and Fernando, 2015). Rehabilitation, for example, after a stroke, requires the controlled motion of limbs; pressure sensors attached to the corresponding muscles monitor their activities and behavior (Lukowicz et al., 2006). To prevent pressure sores (decubitus), pressure sensors can supervise vulnerable body regions, allowing the timely feedback to the wearer (Axisa et al., 2005). Generally these sensors are made with a coating of composite comprising CP, PDMS, and silicone in a knitted fabric (Wang et al., 2014).

Tactile sensing applications are usually used in capacitive fabrics. Their electrical behavior depends on the material sensing composition and on the read-out circuit used. Generally, fabric capacitors consist of two different electrode plates separated by dielectric material, as shown in Fig. 4.14 (Guo et al., 2014). The conductive part, that is, the electrode, can be sewn (Holleczek et al., 2010), woven (Takamatsu et al., 2012), embroidered (Takamatsu et al., 2012) in case of conductive thread, or can be painted, if the conductive material is an ink (Sergio et al., 2003). The dielectrics materials are usually synthetic foams, fabric spacers, or polymers. In principle, a pressure applied on the sensing part generates capacitance variations or a changing of the piezoelectric resonance frequency. For capacitive sensors, a change in parasitic capacitance and resistance can be compensated for by the electronics, therefore the wiring has a marginal influence on the sensed signal. Moreover, the sensor capacitors can be placed in arrays achieving a matrix and distributed pressure sensor. For instance, Meyer et al. (2006)

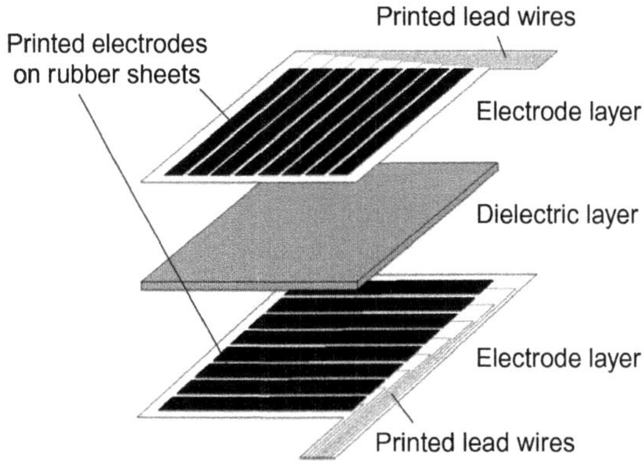

Fig. 4.14 The electrodes on the two-electrode layers are oriented orthogonally and each crossing point of the two perpendicular electrode arrays makes up a capacitive sensor cell on the sheet.
From Guo, S., et al., 2014. A two-ply polymer-based flexible tactile sensor sheet using electric capacitance. Sensors 14 (2), 2225–2238.

integrated a matrix with several capacitive pressure sensors into a garment to detect muscle activity on the upper arm. Xu et al. (2013) instead used a pressure sensor array to analyze the sitting posture for clinical purposes.

Adopting a resistive approach, the pressure sensors, made by a grid of intersecting embroidered conductive yarns, change the contact resistance of intersecting yarns when a pressure is applied. The same result can be achieved using conductive elastic yarns (Wang et al., 2011). They are integrated in a "textile chamber" in which changing the geometry of the sensing material after pressure causes it to respond with an electrical resistance variation. In term of pressure, the sensitivity obtained is about 3 Pa with a pressure range of 0–2 MPa. Fig. 4.15 depicts the working principle in which the teeth geometry of the top and bottom layers deforms the sensing fabric, changing its intrinsic electrical properties.

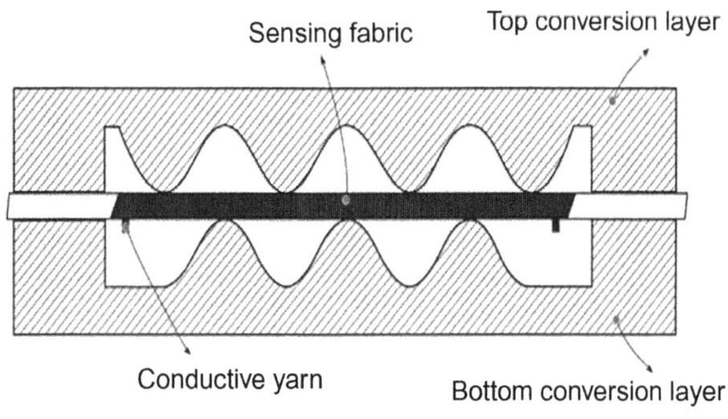

Fig. 4.15 Tooth-structured resistive fabric pressure sensor.
From Wang, Y., et al., 2011. Novel fabric pressure sensors: design, fabrication, and characterization. Smart Mater. Struct. 20 (6), 65015.

This kind of sensor can be manufactured at different levels, that is, fiber, yarn, or coating. In particular, foams can be pressure sensitive thanks to an additional surface coating realized by means of conducting inks which can change their conductance during a contact event (Brady et al., 2005).

The team of the Center for Micro-Bio Robotics of Istituto Italiano di Tecnologia (Pisa, Italy) produced a composite capacitive three-axial sensor based entirely on commercial conductive fabrics, demonstrating its high compliance and stability under manipulation (Viry et al., 2014).

Characterization: Surface morphologies were examined using a JEOL JSM-7001F field emission SEM and the FTIR spectrum was measured with a Bruker Vertex 70 FT-IR spectrometer. The weight fraction of silver in the conductive fiber was evaluated using a thermogravimetry analysis (Q50, TA Instruments) and a mechanical strength test was conducted using USTER TENSORAPID 4. Capacitance measurements were conducted at 300kHz frequency with a 1V AC signal using an Agilent E4980A, Precision 3 LCR Meter. The force and its frequency were mainly controlled by a universal manipulator of Teraleader with 0.01N resolution.

Lee et al. (2015) described a textile-based pressure sensor with high sensitivity (0.21–1 kPa), stability (against 3000 bending tests), a fast response, and a relaxation time provided with a universal manipulator of Teraleader with 0.01N resolution coupled with an Agilent E4980A (300kHz, 1V AC signal). The textile was based on highly conductive fibers coated (0.00015Ω/m) with dielectric rubber materials. The conductive fibers were fabricated by coating poly (styrene-butadiene-styrene) (SBS) polymer on the surface of poly(*p*-phenylene terephthalamide) (Kevlar) fiber, followed by converting a huge amount of silver (Ag) ions (80wt%) into Ag NPs directly in the SBS polymer (Fig. 4.16).

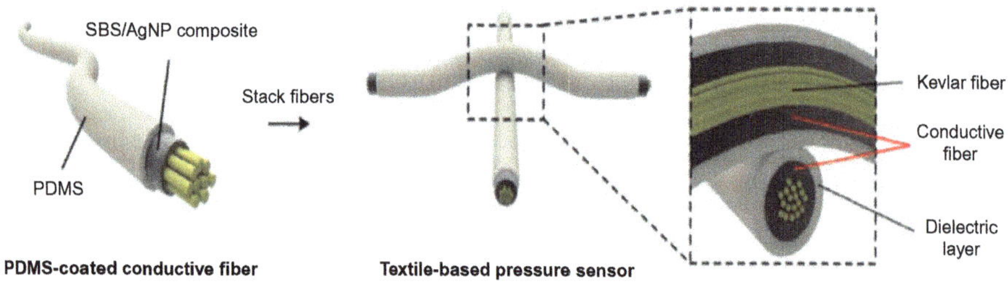

Fig. 4.16 Schematic illustration of the fabrication of the pressure sensor from a single sensing fiber to their implementation.
From Lee, J., et al., 2015. Conductive fiber-based ultrasensitive textile pressure sensor for wearable electronics. Adv. Mater. 27 (15), 2433–2439.

4.2.5.2 Strain sensors

Fabric strain sensors are textile structures that are able to have an electrical response following a mechanical strain. It is possible to design fabric strain sensors at different levels of the structure hierarchy: overlapping of sensing fibers with nonsensing fibers (Araki et al., 2012), adding sensing yarn into fabric (Huang et al., 2008) or coating the fabric with a strain sensor material (Calvert et al., 2008). As occurs for pressure sensors, an ad hoc read-out circuit has to be designed.

In term of applications, wearable stretch sensors are mainly used for sensing and monitoring body parameters, as the textile is in contact with the skin over a large body area. This means that monitoring can take place at several locations on the body (Pacelli et al., 2006). A specific structure of textile sensors can be integrated fibers featuring piezoresistive (Huang et al., 2008), piezoelectric (Nilsson et al., 2013), or piezo capacitive (Koschan et al., 2006) properties, enabling their use as strain or deformation sensor.

According to the type of technologies used, conductive thread knitted into a fabric can be strained up to 65% (Wijesiriwardana et al., 2003). A fabric resistive strain gauge is made using an electroconductive elastomeric fiber (eg, carbon filled polymeric fibers) and nonconductive base fiber (which could be an elastomer). The base structure is formed using the nonconductive fiber and the conductive fiber is laid in the course direction of the base structure. This configuration allows an electrical resistance variation due to the stretching of the fabric. When the strain gauge is stretched in Y (courses) direction by using planar loads, the electrical resistance increases. This is due to the piezoresistive behavior of the composite elastomer (Taya et al., 1998). However, when the strain gauge was stretched in the X (wales) direction, an insignificant variation of the electrical resistance was observed due to the loss of curvature at bending regions. To evaluate this feature, a fiber meshed strain transducer was clamped between the crossbars of a tensile tester (INSTRON 5500) and the upper crossbar was programmed to carry out a dynamic extension of the transducer.

On the other hand, Zhang et al. (2006) developed a strain sensor with carbon fibers knitted reaching a maximum deformation of 30% under a high range of temperature (from 50°C to 200°C), considering that generally the strain gauges are very sensitive to temperature changes and hence require electronic compensation. The carbon fabrics were fabricated by carbonizing the stabilized carbon fabric at the maximum temperature of 1000°C in the inert atmosphere (nitrogen). The diameter of the fabric was about 5 mm and had 12 loops. The tubular structure has more contacting points, and larger internal frictions caused by contacting yarns in comparison to the single warped one. This allows restoration of the original configuration after elongation (see Fig. 4.17).

The experimental setup for the strain and electrical resistance measurements used simultaneously an Instron Universal Material Tester 4466 and a Keithley Multimeter. The researchers adopted a cycle test with a crosshead speed to a given maximum strain.

Stretchsense® developed a soft fabric stretch that transmits information via Bluetooth® about how much they are being stretched thanks to the electronic embedded at the end of the sensing strip (Fig. 4.18).

The Ohmatex Company weaves a very fine constantan (copper–nickel alloy consisting of 55% copper and 45% nickel, Davis et al., 2001) wire into a textile ribbon. This has been used in the Edema medical stocking for measuring changes in leg volume for heart failure patients. This textile ribbon can also be used to measure vibration in large structures such as wind turbine blades (Ohmatex, 2015).

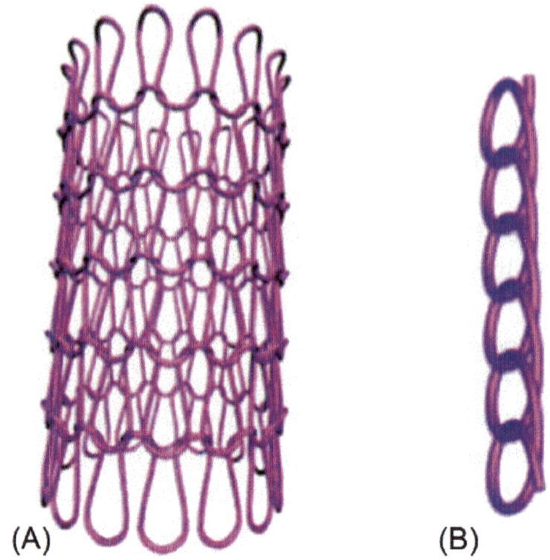

Fig. 4.17 (A) Tubular and (B) single wrap structure.
From Zhang, H., et al., 2006. Conductive knitted fabric as large-strain gauge under high temperature. Sensors Actuators A: Phys. 126 (1), 129–140.

Fig. 4.18 Soft fabric stretch made by Stretchsense®.
From Stretch Sense™.

4.2.5.3 Chemical and gas sensors

Fabrics capable of reacting following a chemical variation can be used as chemical and gas sensors. Relevant applications of these sensors can be found in safety equipment and environmental quality analysis (eg, the presence of harmful gases or water pollutants). To obtain these features, miniaturized chemical/gas sensors can be integrated into a fabric structure by stitching or sewing them. Otherwise, using a polymer coating such as PPy or PANI onto PET can fabricate toxic gas sensors (Hong et al., 2004). Using this method, the fabric sensors are sensitive to the chemical environment variation with a shift of the electric resistance. The specific chemical variation monitored depends on the properties of the sensing film exposed. For instance, PEDOT nanotubes are suitable for monitoring concentration levels of nitric oxide (NO) (Lu et al., 2008). Fig. 4.22 shows two different works about humidity sensors, and in particular, Fig. 4.19A shows a quasi-linear variation of the electric resistance of PEDOT-PSS coated fibers (Windmiller and Wang, 2013) and Fig. 4.19B shows a capacitance variation versus humidity using different sensitive polymeric ink thicknesses (Daoud et al., 2005).

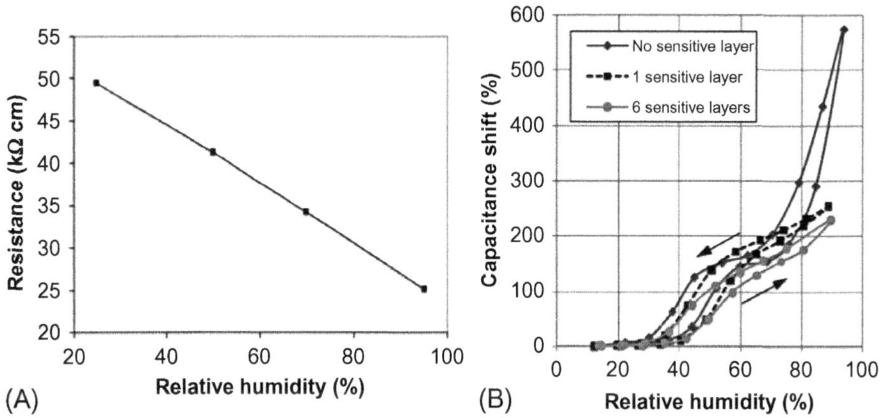

Fig. 4.19 (A) Electric resistance trend from 25% to near humidity saturation. (B) Electric capacitive variation using different sensitive ink layers thicknesses and unmodified polyimide substrate (no sensitive layer).
From Daoud, W.A., Xin, J.H., Szeto, Y.S., 2005. Polyethylenedioxythiophene coatings for humidity, temperature and strain sensing polyamide fibers. Sensors Actuators B: Chem. 109 (2), 329–333.

In terms of biosensing, recent insights into novel fabrication methodologies and electrochemical techniques have shown chemical sensors able to augment conventional physical measurements (ie, heart rate, EEG, ECG, etc.) (Windmiller and Wang, 2013).

BIOTEX, an EU-funded project, aims at developing dedicated biochemical-sensing techniques compatible with integration into textiles. The goal was to monitor the body fluids via sensor distributed on a textile substrate and performing biochemical measurements. The approach involves the development of sensing patches adapted to different body fluids and biological species to be monitored, where the textile itself is the sensor (Coyle et al., 2010).

The researchers of the National Centre for Sensor Research, Dublin City University (Ireland), developed chemical sensors able to measure and analyze sweat in real time on the body. They designed a microchip version of the platform to measure changes in the pH of sweat. The color change of the pH sensitive fabric was detected by placing a surface mount LED and photodiode module on either side of the chip, aligned with the pH sensitive fabric. The final device (180 μm thick) is flexible and can adapt to the body (Benito-Lopez et al., 2009).

4.2.6 Textile energy harvesting systems

Generally, e-textile technology hosts several electronic components, such as sensors, read-out circuits and embedded communication systems. All of them need to be powered, and batteries, or external power supply links, are a limitation for a wearable system. Mechanical stability, bulkiness, and lifetime are only some negative aspects of using an external energy source (Nishide, 2008). For these reasons, the wearable systems should be capable of harvesting energy dissipated by the human body, as motion, muscle stretching, thermal flow or biochemical variations.

Our body is a perfect energy source for wearable devices, and in particular, some daily actions, such as body motion, can generate enough energy to power sensors and/or wearable electronic components. Table 4.2 shows an average of the available power for bodily activities of human beings (Wang, 2008). Wearable electronic devices designed in micro (MEMS) and nanoscale (NEMS) may be powered by the human body, because the energy required is very low. For instance, Infineon is currently trying to recover energy by body movements to feed Mp3 players integrated in a jacket using piezoelectric materials (Infineon, 2015).

The piezoelectric effect is a feasible technique for converting mechanical energy from vibrations to electricity (Nilsson et al., 2014) reaching about 0.7 mW of power from the fibers in the textile. This effect is usefully exploitable with polymeric piezoelectric materials, such as polyvinylidene fluoride and its copolymer, because they are naturally flexible, lightweight, biocompatible and suitable for energy harvesting (Cha et al., 2011). The main issue with the human body interaction is the typical low frequency of mechanical motion (<10 Hz), leading to a low power harvesting.

On the other hand, the body temperature is another exploitable energy source. Feinaeugle et al. (2013) used a fiber-based thermoelectric power generator fabricated by evaporating thin Ni–Ag films on flexible textile substrates and they demonstrated a maximum power harvesting of 2 nW. Semiconductor nanowires exhibit promise for thermoelectric properties, but new, advanced fabrication techniques need to be developed.

Another technology is represented by solar clothing, in which the solar energy is harvested through a new generation of flexible solar cells (Lee et al., 2013). Integration of flexible solar cells into clothing can provide power for portable electronic devices.

Table 4.2 Available power generated daily by human body activities for a 68 kg adult (Wang, 2008)

Energy source	Available power
Footfalls	67 W
Arm motion	60 W
Body heat	2.4–4.8 W
Exhalation	1 W
Blood pressure	0.93 W
Breathing band	0.83 W
Finger motion	6.9–19 mW

Nowadays, the energy demand of portable devices is low enough that clothing-integrated solar cells can power most mobile electronics (Schubert and Werner, 2006).

Enabling textile energy storage requires that the textiles have good electrical conductivity. Flexible energy-storage devices have attracted attention in recent years due to their promising integrability into stretchable and wearable electronics (Hu et al., 2012a). In particular, supercapacitors are of significant interest as energy-storage devices associated with their high power density, long cycling life, and short charging time (Aegerter and Mennig, 2004). Recently, researchers developed a solution-based technique to convert cotton textiles into conductive textiles by coating the cellulose fibers with CNTs (Hu et al., 2010) or graphene (Yu et al., 2011) thin films. The three-dimensional (3D) high-surface-area characteristic of such textiles facilitates the access of electrolytes, enabling high electrochemical performance of textile supercapacitors. Bao and Li (2012) fabricated highly conductive and flexible activated carbon textiles (ACTs) by direct conversion of cotton textiles through:

1. dipping the cotton into 1 M NaF solution;
2. a drying process at 120°C for 3 h;
3. a curing process annealing the dried textiles in a horizontal tubular furnace at 800–1000°C for 1 h under vacuum and inert atmosphere conditions.

SEM analyses showed the textiles consisting of interwoven cellulose fibers with diameters ranging from 5 to 10 μm and a highly conductive under folding condition (sheet resistance ~1–2 Ω/m^2) measured via a Keysight Digital Multimeter. Furthermore, loading the fabric with MnO_2, cyclic voltammetry (CV) tests, a type of potentiodynamic electrochemical measurement, showed an improvement of the capacitance performance with a threefold increase in specific capacitance compared to that of ACTs. At the scan rate of 2 mV/s, the specific capacitance achieved was 269.5 F/g comparable with solution-processed graphene/MnO_2 textiles (Yu et al., 2011).

Jost et al. (2013) have developed textile supercapacitors based on knitted carbon fibers and activated carbon ink with capacitances around 0.51 F/cm at 10 mV/s. They used a solid electrolyte in place of a liquid electrolyte to reduce any possible leakage. The researchers also demonstrated the performance of the device when bent from 90 to 180 degree (Fig. 4.20) showing a small leakage of capacitance after a bending test of around 20%, typically a relevant issue for a supercapacitor. All experiments were carried out using a Biologic VMP3 potentiostat galvanostat (BioLogic, USA). Devices were tested in a two-electrode symmetric setup and subjected to CV,

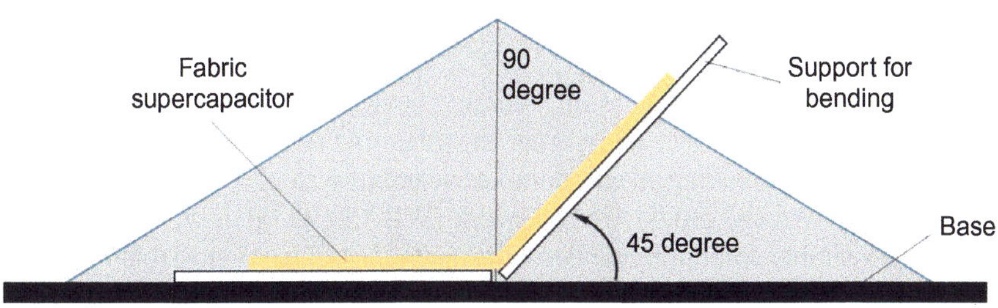

Fig. 4.20 Bending test experimental setup.

galvanostatic cycling, and electrochemical impedance spectroscopy measurements, following procedures fully described in previous studies. All devices were subjected to 100 pre-cycles at $20\,mV\,s^{-1}$, and values of capacitance and ESR were taken from tests after cycling.

They also tested the textile supercapacitor under 50% stretching and the result was a small loss in capacitance due to the breaking of the conductive network between carbon particles when changing the dimensions of the device.

On the other hand, compared with supercapacitors, Li-ion batteries have much higher energy density per weight. The concept of 3D battery electrodes has been used to enhance the capacitance (Long et al., 2004). In recent years, Hu et al. (2011) developed a Li-ion battery based on the conductive textiles in which the electrode materials are loaded into the 3D pores of conductive textiles. They found the stable potential range of conductive textiles in organic electrolytes, and effectively demonstrated a working Li-ion battery with a mass loading of ~168 mg/cm^2 and a thickness of ~600 μm. The textile conductor (sheet resistance of 8 Ω/sq.) is fabricated using a plain polyester fiber textile and well-dispersed single walled CNT ink in water with the addition of 1% sodium dodecylbenzenesulfonate. The conductive textile was found to be electrochemically stable in the voltage window of ~0.9–3.7 V. Either a Bio-Logic VMP3 battery tester or an MTI battery analyzer were used to perform the textile battery. The full cell was assembled and first cycled between 1.4 and 2.6 V for five cycles before the self-discharge test. The batteries were charged to 2.6 V and then disconnected for a certain period of time before a discharge full down to 1.4 V.

In a project funded by the Engineering and Physical Sciences Research Council (EPSRC), Researchers at the University of Southampton's School of Electronics and Computer Science (ECS) are developing an energy harvesting film in textiles using rapid printing processes and active printed inks (Wei et al., 2013).

Instead, Thermotron of UNITIKA Co. is a particular fabric able to convert sunlight into thermal energy while storing heat without wasting it. Inside the Thermotron there are microparticles of zirconium carbide which allow the fabric to absorb and filter sunlight. The inner layer of the fabric withholds the heat generated and prevents it from becoming lost, thus providing a salutary effect on the human body (Unitika LTD, 2015).

Finally, Cetemmsa Technological Centre is overseeing research projects in the use of sensors in sportswear and accessories. It is working on integrated power sources for added electronic functionality such as organic photovoltaic, as part of the EU-funded Dephotex project (Cetemmsa Technology Centre, 2015).

4.2.7 *Wearable antennas*

In recent years, the rapid progress on wearable devices has induced a significant development of wearable antennas, in order to stream the *body-data* information of the wearable sensors and making an autonomous wearable system. This technology is represented by the Wireless Body Area Network (WBAN), in other words, the wireless network of wearable computing devices. The WBAN has many possible applications in the medical field, where the sensors are in direct contact with human body and they stream physiological data in real time, for instance, to the internet. This methodology allows the access of patient data by medical professionals independent of the patient location (Yuce and Khan, 2011).

Wearable antennas, based on conductive textiles, exploit new flexible and smart structures without affecting the native textile properties (Giddens et al., 2012). To achieve this result, conductive textiles such as *Zelt, Flectron* and pure copper polyester fabrics are typically used as the radiating element, while nonconductive textiles are used as substrates (Rais et al., 2009). The geometry of an antenna developed for body wearable applications is shown in Fig. 4.21.

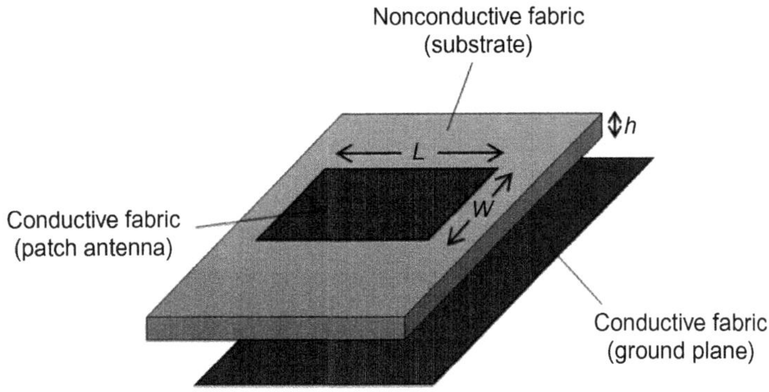

Fig. 4.21 Wearable antenna comprising a thin conductive fabric on an insulating fabric substrate backed by a ground plane.

Generally, body wearable antennas operate at frequencies ranging from 500 MHz to 5 GHz, allowing them to be physically small in construction and relatively simple to wear unobtrusively on the body. However, high frequency bands (>1 GHz) tend to have a very limited range and low wave propagation. The advantages of a high frequency antenna are that they are lightweight, small, and structurally strong (Matthews and Tittensor, 2014). On the other hand, using, for example, a wide band as an FM broadcast band (about 81–130 MHz), a wearable FM antenna should be designed without suffering from the detuning caused by the human body (Roh et al., 2010).

A microstrip patch is a representative candidate for a wearable integration, because it can be thin, lightweight, low maintenance, robust, and easily integrated into a garment and coupled with RF circuits (Wang et al., 2012). Moreover, the conductive textile used for antenna purposes has to have a low and stable electrical resistivity (<1 Ω/sq.) to minimize losses (Locher et al., 2006). Several properties of the materials can influence the behavior of the antenna properties. For instance, the permittivity and the thickness of the substrate change the bandwidth and the efficiency of a planar microstrip antenna (Liu et al., 2011). In general, fabrics present a complex structure, in term of density of fibers and hence air volume and size of the pores, which allow a very low dielectric constant with a reduction of the surface wave losses and an increase of the impedance bandwidth.

Dielectric properties of the nonconductive material play an important role. A wrong choice of material would compromise the performance of the antenna. The dielectric properties are affected by several physical parameters such as: temperature, moisture content, surface roughness, purity, and homogeneity of the material and signal frequency. The permittivity (ε) is a constitutive parameter of dielectrics and usually it is expressed by Eq. (5.2), where ε_0 is the permittivity of vacuum, 8.854 × 10^{-12} F/m (Baker-Jarvis et al., 2010) and ε_r can be expressed applied and the relative displacement.

$$\varepsilon_r : \varepsilon = \varepsilon_0 \cdot \varepsilon_r = \varepsilon_0 \cdot \left(\varepsilon_r' - \mathrm{j} \cdot \varepsilon_r''\right) \tag{5.2}$$

With a low dielectric constant, the surface wave losses are reduced and they are tied to guided wave propagation within substrate. Hence a low ε_r increases the spatial waves and the impedance bandwidth of the antenna (Salvado et al., 2012).

Generally, fabric dielectric properties are represented from the dielectric constant ε_r' and the ratio of the imaginary and to the real part, $\delta = \varepsilon_r'' / \varepsilon_r'$ (Stoppa and Chiolerio, 2014).

In addition, textile structures are composed of fibers, and each fiber can absorb a minimum quantity of water from the environmental moisture, increasing the dielectric constant. To counteract this effect, the antenna dielectric substrate is covered or treated superficially in order to protect the textile from the humidity or varying climatic condition (Hertleer et al., 2009).

Another issue regards the movement of the body that can change the geometry of the antenna and affect the antenna performance. In wearable systems, it is very difficult to keep the antenna flat all the time, especially when the antenna is made of textile materials. Moreover, the wearable antenna is bent frequently due to human body movements changing the electromagnetic properties of the antenna and affecting the data transmission. Sankaralingam and Gupta (2010) investigated the antenna's performance characteristics under bent conditions and they observed that due to bending, the resonant length of the antenna changed and hence, a shift of the resonant frequency occurred too. The tests were carried out by bending the polyester antenna around curved surfaces of cylindrical PVC pipes with different radii: 50.8, 63.5, 76.2, and 88.9 mm. These dimensions are comparable with different parts of the human body like an arm, leg, and shoulder. However, considering other parameters (input impedance bandwidth and input-reflection coefficient at the resonant frequency) there were no relevant changes during bending at different radii, as shown in Fig. 4.22. It is evident that these textile microstrip patch antennas may eventually replace patch antennas on standard printed circuit board substrates for various applications.

From a conductive fabric point of view, for a proper antenna design, the conductivity of the fabric (σ) is a relevant parameter. The conductivity is expressed in Siemens per meter (S/m) by Eq. (5.3):

$$\sigma = 1 / \left(\rho_s \cdot t\right) \tag{5.3}$$

where ρ_s is the surface resistivity and t is the thickness of the fabric. As descripted previously, at the microscale, the textile structure presents several discontinuities and these affect the current flow and the relative σ value (Ouyang et al., 2005).

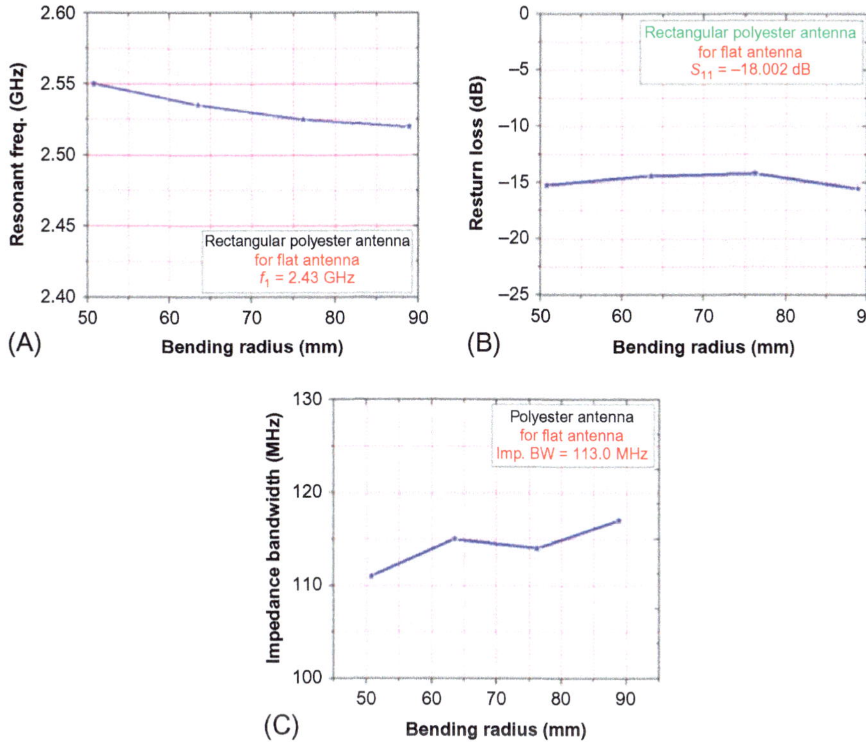

Fig. 4.22 Three different characterizations: (A) resonant frequency, (B) return loss, and (C) impedance bandwidth with different levels of bending radius using a textile antenna based on polyester.
From Sankaralingam, S., Gupta, B., 2010. Development of textile antennas for body wearable applications and investigations on their performance under bent conditions. Prog. Electromagn. Res. B 22, 53–71.

Research of Tampere University of Technology developed a robust textile antenna able to operate in harsh environmental conditions. They manufactured an antenna with different functional textile layers and they improved the performances against moisture and others environmental factors (Lilja et al., 2012).

Monarch Antenna Inc. developed a wearable antenna for soldiers using a flexible substrate worn under a vest. The antenna operates in the unlicensed ISM band of 2.4–2.48 GHz and is currently being redesigned to print on a flexible polymer substrate. Moreover, they collaborated with NASA for developing a multi-beam adaptive wearable antenna for astronauts based on Monarch's patented SSA technology (Monarch Antenna Inc., 2015).

Pharad company is a supplier of wearable antenna products. Incorporating Pharad's patented Flextenna®, they developed a flexible antenna technology for wearable applications such as a solution for first responders, soldiers, marines, and security/intelligence personnel operating covertly. Pharad offers various mounting configurations, and standard connector options allow these antennas to easily connect to most radios. The wearable antennas developed by Pharad have different bandwidths from 2–30 MHz to 3–10 GHz according to the applications (Pharad, 2015).

4.3 Conclusions

Electronic textiles describe the convergence of electronics and textiles into fabrics which are able to sense, compute, communicate, and actuate. This chapter has provided a review of the current state-of-the-art e-textiles from a manufacturing and performance assessment point of view. It has been relevant to highlight different approaches at different levels of textile integration, which show a common aim: to develop the most efficient and high performance electronic-textile structure. In relation with the method adopted to achieve fibers, yarns, or fabrics with electrical or sensing properties, the raw materials choice plays an important role. Current advances in textile technologies, new materials, nanotechnology, and miniaturized electronics are making wearable systems more feasible, nevertheless, the final key factor for user acceptance of wearable devices is the fit comfort. Overall, the desired outcome is to seek proper materials capable of interfacing with the textile structures, adding electrical/sensing features without affecting the original properties of the fabric, such as flexibility, wearability, comfort, and washability. Several issues are still present, including performance loss upon bending/stretching/washing, that still require a huge contribution from the already recalled fields, in particular, materials science and nanotechnology, to be solved.

The vision of wearable computing describes future electronic systems as an integral part of our everyday clothing serving as intelligent personal assistants. Therefore, such wearable sensors must maintain their sensing capabilities under the demands of normal wear, which can impose severe mechanical deformation of the underlying garment/substrate.

The potential applications are several and e-textile technologies have a direct link with human body monitoring. "Clothes are our own personal house" (Van Langenhove, 2007) and there are no other devices continuously in contact with the human body like garments. In this way, the combination of smart materials, wearable electronics, and sensors with fabrics opens the doors to many applications, especially in the clinical monitoring field. Textiles and clothes can be produced on fast and productive machinery at a reasonable cost, and the health and beauty industry is also taking advantage of these innovations.

Finally, the vision behind wearable computing foresees future electronic systems to be an integral part of our everyday outfits and the knowledge of circuit design, smart materials, micro electronics, and chemistry must be integrated to understand the new generation of textile fabrication.

References

Aegerter, M.A., Mennig, M., 2004. Sol–Gel Technologies for Glass Producers and Users. Springer Science & Business Media, New York.

Alzaidi, A., Zhang, L., Bajwa, H., 2012. Smart textiles based wireless ECG system. In: Systems, Applications and Technology Conference (LISAT), 2012 IEEE Long Island, pp. 1–5.

Ansar, M., Xinwei, W., Chouwei, Z., 2011. Modeling strategies of 3D woven composites: a review. Compos. Struct. 93 (8), 1947–1963.

Araki, T., et al., 2011. Printable and stretchable conductive wirings comprising silver flakes and elastomers. IEEE Electron Device Lett. 32 (10), 1424–1426.

Araki, T., et al., 2012. Experimental investigation of surface identification ability of a low-profile fabric tactile sensor. In: 2012 IEEE/RSJ International Conference on Intelligent Robots and Systems (IROS), pp. 4497–4504.

Ashruf, C.M.A., 2002. Thin flexible pressure sensors. Sens. Rev. 22 (4), 322–327. Available at: http://www.emeraldinsight.com/journals.htm?issn=0260-2288&volume=22&issue=4&articleid=876363&show=html (accessed 20.01.14).

Atwa, Y., Maheshwari, N., Goldthorpe, I.A., 2015. Silver nanowire coated threads for electrically conductive textiles. J. Mater. Chem. C3, 3908.

Axisa, F., et al., 2005. Flexible technologies and smart clothing for citizen medicine, home healthcare, and disease prevention. IEEE Trans. Inf. Technol. Biomed. 9 (3), 325–336.

Baik, W., et al., 2009. Synthesis of highly conductive poly (3,4-ethylenedioxythiophene) fiber by simple chemical polymerization. Synth. Met. 159 (13), 1244–1246.

Baker-Jarvis, J., Janezic, M., DeGroot, D.C., 2010. High-frequency dielectric measurements. IEEE Instrum. Meas. Mag. 13 (2), 24–31.

Baltex, 2014. Available at: http://www.baltex.co.uk/home/ (accessed 14.05.15).

Bao, L., Li, X., 2012. Towards textile energy storage from cotton T-shirts. Adv. Mater. 24 (24), 3246–3252.

Bashir, T., 2013. Conjugated Polymer-Based Conductive Fibers for Smart Textile Applications. University of Boras, Sweden.

Benatti, S., Benini, L., Farella, E., 2014. Towards EMG control interface for smart garments. In: Proceedings of the 2014 ACM International Symposium on Wearable Computers: Adjunct Program, pp. 163–170.

Benito-Lopez, F., et al., 2009. Pump less wearable microfluidic device for real time pH sweat monitoring. Proc. Chem. 1 (1), 1103–1106.

Bigg Battelle, D.M., 1977. Conductive polymeric compositions. Polym. Eng. Sci. 17 (12), 842–847.

Black, S., 2007. Trends in smart medical textiles. In: Smart Textiles for Medicine and Healthcare—Materials, Systems and Application. Woodhead Publishing Ltd, Cambridge, pp. 3–26. Available at: http://www.woodheadpublishing.com/en/book.aspx?bookID=888.

Blayo, A., Pineaux, B., 2005. Printing processes and their potential for RFID printing. In: Proceedings of the 2005 Joint Conference on Smart Objects and Ambient Intelligence: Innovative Context-Aware Services: Usages and Technologies, pp. 27–30.

Blum, F., 2000. Method for producing an electrically conductive yarn, the electrically conductive yarn and use of the electrically conductive yarn. US Patent Number 6032450, March 7. Filed on June 30, 1997.

Bocchini, S., Chiolerio, A., Porro, S., Accardo, D., Garino, N., Bejtka, K., Perrone, D., Pirri, C.F., 2013. Synthesis of polyaniline-based inks, doping thereof and test device printing towards electronic applications. J. Mater. Chem. C1, 5101–5109.

Brady, S., Diamond, D., Lau, K.-T., 2005. Inherently conducting polymer modified polyurethane smart foam for pressure sensing. Sensors Actuators A Phys. 119 (2), 398–404.

Calvert, P., et al., 2008. Conducting polymer and conducting composite strain sensors on textiles. Mol. Cryst. Liq. Cryst. 484 (1), 291–657.

Camarchia, V., et al., 2014. Demonstration of inkjet-printed silver nanoparticle microstrip lines on alumina for RF power modules. Org. Electron. 15 (1), 91–98.

Canshielding, 2015. JP Canshielding Corporation. Available at: www.canshielding.com (accessed 01.06.15).

Castano, L.M., Flatau, A.B., 2014. Smart fabric sensors and e-textile technologies: a review. Smart Mater. Struct. 053001 (23), 1–27.
Cetemmsa Technology Centre, 2015. Cetemmsa. Available at: http://www.cetemmsa.com/ (accessed 10.05.15).
Cha, S., et al., 2011. Porous PVDF as effective sonic wave driven nanogenerators. Nano Lett. 11 (12), 5142–5147.
Chauraya, A., et al., 2013. Inkjet printed dipole antennas on textiles for wearable communications. IET Microw. Antenna Propag. 7 (9), 760–767.
Chiolerio, A., Cotto, M., Pandolfi, P., Martino, P., Camarchia, V., Pirola, M., Ghione, G., 2012. Ag nanoparticle-based inkjet printed planar transmission lines for RF and microwave applications: considerations on ink composition, nanoparticle size distribution and sintering time. Microelectron. Eng. 97, 8–15.
Cottet, D., et al., 2003. Electrical characterization of textile transmission lines. IEEE Trans. Adv. Pack. 26 (2), 182–190.
Coyle, S., et al., 2010. BIOTEX—biosensing textiles for personalised healthcare management. IEEE Trans. Inf. Technol. Biomed. 14 (2), 364–370.
Dalmas, F., et al., 2006. Carbon nanotube-filled polymer composites. Numerical simulation of electrical conductivity in three-dimensional entangled fibrous networks. Acta Mater. 54 (11), 2923–2931.
Daoud, W.A., Xin, J.H., Szeto, Y.S., 2005. Polyethylenedioxythiophene coatings for humidity, temperature and strain sensing polyamide fibers. Sensors Actuators B Chem. 109 (2), 329–333.
Davis, J.R., et al., 2001. Copper and Copper Alloys. ASM International, Ohio.
De Vries, H., Peerlings, R., 2014. Predicting conducting yarn failure in woven electronic textiles. Microelectron. Reliab. 54 (12), 2956–2960.
Ding, Y., Invernale, M.A., Sotzing, G.A., 2010. Conductivity trends of PEDOT-PSS impregnated fabric and the effect of conductivity on electrochromic textile. ACS Appl. Mater. Interfaces 2 (6), 1588–1593.
Dordevic, Z., 1992. Textile fabric shielding electromagnetic radiation, and clothing made thereof. Available at: https://www.google.com/patents/US5103504.
Dubrovski, P.D., Cebasek, P.F., 2005. Analysis of the mechanical properties of woven and nonwoven fabrics as an integral part of compound fabrics. Fibres Text. East. Eur. 13 (3), 50.
Elektrisola, 2015. Elektrisola Feindraht AG. Available at: http://www.elektrisola.com/ (accessed 15.05.15).
Farboodmanesh, S., et al., 2005. Effect of coating thickness and penetration on shear behavior of coated fabrics. J. Elastom. Plast. 37 (3), 197–227.
Feinaeugle, M., et al., 2013. Fabrication of a thermoelectric generator on a polymer-coated substrate via laser-induced forward transfer of chalcogenide thin films. Smart Mater. Struct. 22 (11), 115023.
Fleury, A., Sugar, M., Chau, T., 2015. E-textiles in clinical rehabilitation: a scoping review. Electronics 4 (1), 173–203. Available at: http://www.mdpi.com/2079-9292/4/1/173/ (accessed 05.05.15).
Fujimori, T., et al., 2011. Enhanced X-ray shielding effects of carbon nanotubes. Mater. Express 1 (4), 273–278.
Giddens, H., et al., 2012. Influence of body proximity on the efficiency of a wearable textile patch antenna. In: 2012 6th European Conference on Antennas and Propagation (EUCAP), pp. 1353–1357.
Grillet, A., et al., 2008. Optical fiber sensors embedded into medical textiles for healthcare monitoring. IEEE Sens. J. 8 (7), 1215–1222.

Guo, L., Berglin, L., 2009. Test and evaluation of textile based stretch sensors. In: AUTEX 2009 World Textile Conference, Izmir, Turkey, p. 8.
Guo, S., et al., 2014. A two-ply polymer-based flexible tactile sensor sheet using electric capacitance. Sensors 14 (2), 2225–2238.
Hamdani, S.T.A., Fernando, A., 2015. The application of a piezo-resistive cardiorespiratory sensor system in an automobile safety belt. Sensors 15, 7742–7753.
Hertleer, C., et al., 2009. Influence of relative humidity on textile antenna performance. Text. Res. J. 80, 177–183.
Holleczek, T., et al., 2010. Textile pressure sensors for sports applications. In: 2010 IEEE Sensors, pp. 732–737.
Hong, K.H., Oh, K.W., Kang, T.J., 2004. Polyaniline–nylon 6 composite fabric for ammonia gas sensor. J. Appl. Polym. Sci. 92 (1), 37–42.
Hu, L., et al., 2010. Stretchable, porous, and conductive energy textiles. Nano Lett. 10 (2), 708–714.
Hu, L., et al., 2011. Lithium-ion textile batteries with large areal mass loading. Adv. Energy Mater. 1 (6), 1012–1017.
Hu, C., et al., 2012a. Graphene microtubings: controlled fabrication and site-specific functionalization. Nano Lett. 12 (11), 5879–5884.
Hu, J., et al., 2012b. A review of stimuli-responsive polymers for smart textile applications. Smart Mater. Struct. 21 (5), 53001.
Huang, C.-T., et al., 2008. A wearable yarn-based piezo-resistive sensor. Sensors Actuators A Phys. 141 (2), 396–403.
Hunt, M.A., et al., 2012. Patterned functional carbon fibers from polyethylene. Adv. Mater. 24 (18), 2386–2389.
Im, S.G., Gleason, K.K., 2007. Systematic control of the electrical conductivity of poly (3,4-ethylenedioxythiophene) via oxidative chemical vapor deposition. Macromolecules 40 (18), 6552–6556.
Industries, T., 2015. Taber Industries. Available at: http://www.taberindustries.com/ (accessed 04.06.15).
Infineon, 2015. Infineon Technologies. Available at: https://www.infineon.com/ (accessed 10.05.15).
Ionescu, C., et al., 2012. Investigations on current capabilities of PEDOT:PSS conductors. In: 2012 35th International Spring Seminar on Electronics Technology (ISSE), pp. 59–64.
Jost, K., et al., 2013. Knitted and screen printed carbon-fiber supercapacitors for applications in wearable electronics. Energ. Environ. Sci. 6 (9), 2698–2705.
Jung, S., et al., 2003. Enabling technologies for disappearing electronics in smart textiles. In: Proceedings of IEEE ISSCC03, vol. 1, pp. 386–387.
Kallmayer, C., Simon, E., 2012. Large area sensor integration in textiles. In: 2012—9th International Multi-Conference on Systems, Signals and Devices.
Karaguzel, B., et al., 2009. Flexible, durable printed electrical circuits. J. Text. Inst. 100 (1), 1–9. Available at: http://dx.doi.org/10.1080/00405000802390147 (accessed 09.01.14).
Kim, H., et al., 2008. A 1.12 mW continuous healthcare monitor chip integrated on a planar fashionable circuit board. In: 2008 IEEE International Solid-State Circuits Conference—Digest of Technical Papers. IEEE, New Jersey, pp. 150–603. Available at: http://ieeexplore.ieee.org/lpdocs/epic03/wrapper.htm?arnumber=4523101 (accessed 20.01.14).
Kim, Y., Kim, H., Yoo, H.-J., 2010. Electrical characterization of screen-printed circuits on the fabric. IEEE Trans. Adv. Pack. 33 (1), 196–205.
Koschan, A., et al., 2006. Design and microfabrication of a hybrid piezoelectric-capacitive tactile sensor. Sens. Rev. 26 (3), 186–192.

Laukhina, E., et al., 2014. Conductive fabric responding to extremely small temperature changes. Proc. Eng. 87, 144–147.

Lee, S., et al., 2010. Arm-band type textile-MP3 player with multi-layer planar fashionable circuit board (P-FCB) techniques. In: ISWC, pp. 1–7.

Lee, Y.-H., et al., 2013. Wearable textile battery rechargeable by solar energy. Nano Lett. 13 (11), 5753–5761.

Lee, J., et al., 2015. Conductive fiber-based ultrasensitive textile pressure sensor for wearable electronics. Adv. Mater. 27 (15), 2433–2439.

LessEMF, 2015. LessEMF. Available at: http://www.lessemf.com/fabric.html.

Lilja, J., et al., 2012. Design and manufacturing of robust textile antennas for harsh environments. IEEE Trans. Antennas Propag. 60 (9), 4130–4140.

Liu, N., et al., 2011. Electromagnetic properties of electro-textiles for wearable antennas applications. Front. Electr. Electron. Eng. China 6 (4), 563–566.

Locher, I., 2006. Technologies for System-on-Textile Integration. Available at: http://e-collection.library.ethz.ch/view/eth:28457 (accessed 20.01.14).

Locher, I., et al., 2006. Design and characterization of purely textile patch antennas. IEEE Trans. Adv. Pack. 29 (4), 777–788.

Löfhede, J., Seoane, F., Thordstein, M., 2012. Textile electrodes for EEG recording—a pilot study. Sensors 12 (12), 16907–16919.

Long, J.W., et al., 2004. Three-dimensional battery architectures. Chem. Rev. 104 (10), 4463–4492.

Lu, H.-H., et al., 2008. NO gas sensor of PEDOT: PSS nanowires by using direct patterning DPN. In: Engineering in Medicine and Biology Society, 2008. EMBS 2008. 30th Annual International Conference of the IEEE, pp. 3208–3211.

Lukowicz, P., et al., 2006. Detecting and interpreting muscle activity with wearable force sensors. In: Pervasive Computing. Springer, Berlin, Heidelberg, pp. 101–116.

Materials, A.S., 2012. Annual Book of ASTM Standards *Section 7* Textiles, Vol. 07.01-07.02: Textiles—Yarns, Fabrics, and General Test Methods. American Society for Testing and Materials, West Conshohocken. Available at: https://books.google.co.uk/books?id=8YA9MwEACAAJ.

Matthews, J.C.G., Tittensor, P.J., 2014. Body Wearable Antenna. Available at: https://www.google.co.uk/patents/EP2529446B1?cl=en.

McCann, J., 2013. Smart protective textiles for older people. In: A Smart Textiles for Protection. Woodhead Publishing Ltd, Cambridge, pp. 224–275.

Mcfarland, E.G., et al., 1999. Effects of moisture and fiber type on infrared absorption of fabrics. Text. Res. J. 69, 607–615.

Meoli, D., May-Plumlee, T., 2002. Interactive electronic textile development: a review of technologies. JTATM 2 (2), 1–12.

Merilampi, S., et al., 2010. Analysis of electrically conductive silver ink on stretchable substrates under tensile load. Microelectron. Reliab. 50 (12), 2001–2011.

Meyer, J., Lukowicz, P., Troster, G., 2006. Textile pressure sensor for muscle activity and motion detection. In: 2006 10th IEEE International Symposium on Wearable Computers. IEEE, New Jersey, pp. 69–72. Available at: http://ieeexplore.ieee.org/lpdocs/epic03/wrapper.htm?arnumber=4067729 (accessed 20.01.14).

Monarch Antenna Inc., 2015. Available at: http://www.monarchantenna.com/ (accessed 15.06.15).

Nakad, Z.S., 2003. Architectures for e-Textiles. Virginia Tech, Blacksburg, VA. Available at: https://vtechworks.lib.vt.edu/bitstream/handle/10919/11084/Dissertation.pdf?sequence=1&isAllowed=y.

Needles, H.L., 1981. Handbook of Textile Fibers, Dyes and Finishes. Garland STPM Press, New York.
Nilsson, E., et al., 2013. Poling and characterization of piezoelectric polymer fibers for use in textile sensors. Sensors Actuators A Phys. 201, 477–486.
Nilsson, E., et al., 2014. Energy harvesting from piezoelectric textile fibers. Proc. Eng. 87, 1569–1572.
Nishide, H., Oyaizu, K., 2008. Toward flexible batteries. Science 319 (5864), 737–738.
Numakura, D., 2008. Advanced screen printing "Practical approaches for printable & flexible electronics". In: Microsystems, Packaging, Assembly & Circuits Technology Conference, 2008. IMPACT 2008. 3rd International, pp. 205–208.
Oh, K.W., Park, H.J., Kim, S.H., 2003. Stretchable conductive fabric for electrotherapy. J. Appl. Polym. Sci. 88 (5), 1225–1229.
Ohmatex, 2015. Ohmatex. Available at: http://www.ohmatex.dk/ (accessed 10.05.15).
Omenetto, F., et al., 2011. Silk based biophotonic sensors. US Patent Application Number 13/813,288, December 12, 2013. Filed on July 30, 2011.
Orth, M., 2002. Defining flexibility and sewability in conductive yarns. In: MRS Proceedings, pp. D1–D4.
Ouyang, Y., Karayianni, E., Chappell, W.J., 2005. Effect of fabric patterns on electrotextile patch antennas. In: 2005 IEEE Antennas and Propagation Society International Symposium, pp. 246–249.
Pacelli, M., et al., 2006. Sensing fabrics for monitoring physiological and biomechanical variables: E-textile solutions. In: Proceedings of the 3rd IEEE-EMBS, pp. 1–4.
Parashkov, R., et al., 2005. Large area electronics using printing methods. Proc. IEEE 93 (7), 1321–1329.
Pereira, C.C., Nobrega, R., Borges, C.P., 2000. Spinning process variables and polymer solution effects in the die-swell phenomenon during hollow fiber membranes formation. Braz. J. Chem. Eng. 17 (4–7), 599–606.
Perelshtein, I., et al., 2008. Sonochemical coating of silver nanoparticles on textile fabrics (nylon, polyester and cotton) and their antibacterial activity. Nanotechnology 19 (24), 245705.
Pharad, 2015. Pharad. Available at: http://www.pharad.com/wearable-antennas.html (accessed 15.06.15).
Pomfret, S.J., et al., 2000. Electrical and mechanical properties of polyaniline fibres produced by a one-step wet spinning process. Polymer 41 (6), 2265–2269.
Post, E.R., et al., 2000. E-broidery: design and fabrication of textile-based computing. IBM Syst. J. 39 (3.4), 840–860. Available at: http://dl.acm.org/citation.cfm?id=1011416.1011448 (accessed 14.01.14).
Rais, N.H.M., et al., 2009. A review of wearable antenna. In: Antennas & Propagation Conference, 2009. LAPC 2009, Loughborough, pp. 225–228.
Resistat, 2015. Resistat Fiber Collection. Available at: http://www.resistat.com (accessed 10.06.15).
Roh, J.-S., et al., 2010. Embroidered wearable multiresonant folded dipole antenna for FM reception. IEEE Antenn. Wireless Propag. Lett. 9, 803–806.
Salvado, R., et al., 2012. Textile materials for the design of wearable antennas: a survey. Sensors 12 (11), 15841–15857.
Sankaralingam, S., Gupta, B., 2010. Development of textile antennas for body wearable applications and investigations on their performance under bent conditions. Prog. Electromagn. Res. B 22, 53–71.
Scalisi, R.G., et al., 2015. Inkjet printed flexible electrodes for surface electromyography. Org. Electron. 18, 89–94.

Schneegass, S., et al., 2014. Workshop on smart garments: sensing, actuation, interaction, and applications in garments. In: Proceedings of the 2014 ACM International Symposium on Wearable Computers: Adjunct Program, pp. 225–229.
Schroder, D.K., 2006. Semiconductor Material and Device Characterization. John Wiley & Sons, New Jersey.
Schubert, M.B., Werner, J.H., 2006. Flexible solar cells for clothing. Mater. Today 9 (6), 42–50.
Schwarz, A., et al., 2009. How to equip para-aramide yarns with electro-conductive properties. In: Sixth International Workshop on Wearable and Implantable Body Sensor Networks, 2009. BSN 2009, pp. 278–281.
Schwarz, A., Cuny, L., et al., 2011a. Electrical circuit model of elastic and conductive yarns produced by hollow spindle spinning. Mater. Sci. Technol. 26 (3), 121–127.
Schwarz, A., Kazani, I., et al., 2011b. Electro-conductive and elastic hybrid yarns—the effects of stretching, cyclic straining and washing on their electro-conductive properties. Mater. Des. 32 (8), 4247–4256.
Sen, A.K., 2007. Coated Textiles: Principles and Applications, second ed. CRC Press, Boca Raton.
Sergio, M., et al., 2003. A dynamically reconfigurable monolithic CMOS pressure sensor for smart fabric. IEEE J. Solid-State Circ. 38 (6), 966–975.
Shirakawa, H., et al., 1977. Synthesis of electrically conducting organic polymers: halogen derivatives of polyacetylene, $(CH)_x$. J. Chem. Soc. Chem. Commun. 16, 578–580.
Sinclair, R., 2014. Textiles and Fashion: Materials, Design and Technology. Elsevier, Cambridge.
Sophitex, 2015. Sophitex. Available at: http://www.sophitex.com/ (accessed 10.06.15).
Sprint Metal, 2015. Sprint Metal. Available at: http://sprintmetal.schmolz-bickenbach.com/ (accessed 01.06.15).
Stassi, S., et al., 2014. Nanosized gold and silver spherical, spiky, and multi-branched particles. Handbook of Nanomaterials Properties. Springer, Berlin, Heidelberg, pp. 179–212.
Statex, 2015. Statex. Available at: http://www.statex.de/index.php/en/ (accessed 10.06.15).
Stoppa, M., Chiolerio, A., 2014. Wearable electronics and smart textiles: a critical review. Sensors 14 (7), 11957–11992.
Sülar, V., Okur, A., 2008. Objective evaluation of fabric handle by simple measurement methods. Text. Res. J. 78 (10), 856–868.
Swiss Shield, 2015. Swiss Shield. Available at: http://www.swiss-shield.ch/ (accessed 15.05.15).
Takamatsu, S., et al., 2012. Fabric pressure sensor array fabricated with die-coating and weaving techniques. Sensors Actuators A Phys. 184, 57–63.
Taya, M., Kim, W.J., Ono, K., 1998. Piezoresistivity of a short fiber/elastomer matrix composite. Mech. Mater. 28 (1), 53–59.
Thongruang, W., Spontak, R.J., Balik, C.M., 2002. Correlated electrical conductivity and mechanical property analysis of high-density polyethylene filled with graphite and carbon fiber. Polymer 43 (8), 2279–2286.
Tiberto, P., et al., 2013. Magnetic properties of jet-printer inks containing dispersed magnetite nanoparticles. Eur. Phys. J. B 86 (4), 1–6.
Tibtech, 2015. Tibtech. Available at: http://www.tibtech.com/metal_fiber_composition.php (accessed 11.06.15).
Ugitech, 2015. Ugitech. Available at: http://www.ugitech.com/ (accessed 01.06.15).
Unitika LTD, 2015. Unitika LTD. Available at: http://www.unitika.co.jp/e/ (accessed 15.05.15).
Van Langenhove, L., 2007. Smart Textiles for Medicine and Healthcare: Materials, Systems and Applications. Elsevier, New York.
Van Langenhove, L., Hertleer, C., 2004. Smart clothing: a new life. Int. J. Cloth. Sci. Technol. 16 (1/2), 63–72.

Vatansever, D., et al., 2011. Smart Woven Fabrics in Renewable Energy Generation. INTECH Open Access Publisher, United Kingdom.
Viry, L., et al., 2014. Flexible three-axial force sensor for soft and highly sensitive artificial touch. Adv. Mater. 26 (17), 2659–2664.
Wang, Z.L., 2008. Towards self-powered nanosystems: from nanogenerators to nanopiezotronics. Adv. Funct. Mater. 18 (22), 3553–3567.
Wang, Y., et al., 2011. Novel fabric pressure sensors: design, fabrication, and characterization. Smart Mater. Struct. 20 (6), 65015.
Wang, Z., et al., 2012. Flexible textile antennas for body-worn communication. In: 2012 IEEE International Workshop on Antenna Technology (iWAT), pp. 205–208.
Wang, F., et al., 2014. Flexible pressure sensors for smart protective clothing against impact loading. Smart Mater. Struct. 23 (1), 15001.
Washing, T.-D., 2000. Drying Procedures for Textile Testing. ISO 6330-2000.
Wei, Y., et al., 2013. Screen printing of a capacitive cantilever-based motion sensor on fabric using a novel sacrificial layer process for smart fabric applications. Meas. Sci. Technol. 24 (7), 75104.
Wijesiriwardana, R., Dias, T., Mukhopadhyay, S., 2003. Resistive fibre-meshed transducers. In: 2012 16th International Symposium on Wearable Computers, p. 200.
Windmiller, J.R., Wang, J., 2013. Wearable electrochemical sensors and biosensors: a review. Electroanalysis 25 (1), 29–46.
Winterhalter, C.A., et al., 2005. Development of electronic textiles to support networks, communications, and medical applications in future US Military protective clothing systems. IEEE Trans. Inf. Technol. Biomed. 9 (3), 402–406.
Xu, W., et al., 2013. eCushion: a textile pressure sensor array design and calibration for sitting posture analysis. IEEE Sens. J. 13 (10), 3926–3934.
Xue, P., et al., 2004. Electromechanical behavior of fibers coated with an electrically conductive polymer. Text. Res. J. 74 (10), 929–936.
Yamashita, T., Miyake, K., Itoh, T., 2012. Conductive polymer coated elastomer contact structure for woven electronic textile. In: 2012 IEEE 25th International Conference on Micro Electro Mechanical Systems (MEMS), pp. 408–411.
Yu, G., et al., 2011. Enhancing the supercapacitor performance of graphene/MnO_2 nanostructured electrodes by conductive wrapping. Nano Lett. 11 (10), 4438–4442.
Yuce, M.R., Khan, J., 2011. Wireless Body Area Networks: Technology, Implementation, and Applications. Pan Stanford Publishing, Singapore.
Zeagler, C., et al., 2013. Can I wash it? The effect of washing conductive materials used in making textile based wearable electronic interfaces. In: Proceedings of the 2013 International Symposium on Wearable Computers, pp. 143–144.
Zhang, H., et al., 2006. Conductive knitted fabric as large-strain gauge under high temperature. Sensors Actuators A Phys. 126 (1), 129–140.

Acoustic testing and evaluation of textiles for buildings and office environments

5

X. Qiu
RMIT University, Melbourne, VIC, Australia

This chapter first provides an introduction to the acoustic properties of textiles, which include propagation, absorption, and scattering of sound. The properties can be characterized by various parameters such as flow resistance, transmission loss, absorption coefficient, and scattering coefficient. Test and evaluation methods for obtaining these parameters are discussed. Based on the acoustic properties of the textiles, acoustics designers can make use of textiles in buildings and office environments to optimize sound quality depending on particular requirements.

5.1 Acoustic properties of textiles

A textile is a flexible woven material consisting of a network of natural or artificial fibers, often referred to as thread, which is produced by spinning raw fibers of wool, flax, cotton, or other material to produce long strands (https://en.wikipedia.org/wiki/Textile). Textiles are formed by weaving, knitting, crocheting, knotting, entangling, or felting. The acoustic properties of textiles can be characterized in three categories as propagation, absorption, and scattering, and these in turn can be related to the measurable quantities: (air) flow resistance, acoustic transmission loss, acoustic absorption coefficient, and acoustic scattering coefficient.

5.1.1 Flow resistivity

Most textiles that are used for acoustic purposes show open porosity due to many interconnected pores or voids inside. The acoustic performance of a porous textile is mainly determined by its (air) flow resistivity, which is an intrinsic property of the textile and is a measure of how easily air can enter and pass through a porous textile material (Cox and D'Antonio, 2009). Flow resistivity, also known as static flow resistivity, is related to acoustical properties and plays a critical role in the calculation of many intrinsic acoustic properties of porous textiles, such as the characteristic impedance, the propagation constant, and the sound absorption coefficient. In the S.I. Unit system, flow resistivity is quoted in units of N s m^{-4} and is defined as the unit-thickness specific flow resistance σ (Morfey, 2001),

Performance Testing of Textiles. http://dx.doi.org/10.1016/B978-0-08-100570-5.00006-2

$$\sigma = \frac{R_f}{h} \tag{6.1}$$

where R_f is the specific flow resistance of a uniform textile layer of thickness h and is defined as

$$R_f = \frac{\Delta p\, S}{V_0} \tag{6.2}$$

where Δp is the pressure drop through the layer with a surface area of S under conditions of slow steady flow with a volume flow rate of V_0. The unit of the specific flow resistance is $\mathrm{N\,s\,m^{-3}}$. V_0/S is the volume flow rate per unit surface area at the face of the layer, and is called the face velocity. Here, the label "specific" indicates that R_f is the flow resistance of a sample that has a unit cross sectional area, and is measured transversely to the flow.

Flow resistance is also known as static flow resistance and is the ratio of the pressure drop across a porous element to the volume velocity flowing through it under conditions of steady low speed flow. The flow resistance is almost independent of the volume velocity at low speeds. However, flow resistance is dependent on the acoustic frequency. Dynamic specific flow resistance of a thin (compared to acoustic wavelengths) porous textile layer is the real part of the complex specific flow impedance at a specified frequency, which is defined as the complex ratio of the pressure drop across the layer to the relative face velocity through the layer. When the frequency tends to zero, the dynamic specific flow resistance varies little with frequency, so it is almost equal to the static flow resistance. In the international standard, flow resistance is defined as the real part of the ratio between the pressure drop and the flow velocity through a layer of material of unit thickness (ISO 9053, 1991). Flow resistance characterizes a layer of specified thickness, whereas flow resistivity characterizes a bulk material in terms of resistance per unit thickness.

For fibrous porous materials made from fiberglass and rockwool, with fiber diameters from 1 to 15 μm, their flow resistivity can be calculated by Eq. (6.3) (Bies and Hansen, 2009):

$$\sigma = 27.3\left(\frac{\rho_m}{\rho_f}\right)^{k_1}\frac{\mu}{d^2} \tag{6.3}$$

where ρ_m is the porous material bulk density, ρ_f is the fiber material density, μ is the dynamic gas viscosity ($1.84\times10^{-5}\ \mathrm{kg\,m^{-1}\,s^{-1}}$ for air at 20°C), d is the fiber diameter of the porous material, and the constant $k_1 = 1.53$ for fiberglass and rockwool fibers with diameters between 1 and 15 μm and might have a different value for other fibers with different diameters. It is obvious from Eq. (6.3) that flow resistivity can be increased by decreasing the fiber diameter. For acoustical textiles, the range of values for the static air flow resistivity is approximately from 10^3 to $10^6\ \mathrm{N\,s\,m^{-4}}$.

5.1.2 Propagation

For a sound wave propagating in porous textiles, the porous gas-filled medium is often treated as an equivalent uniform medium for analysis purposes, so a propagation factor $e^{j\omega t-\gamma x}$ can be used to describe the dependence of the propagating wave on time t and the propagation coordinate x (Bies and Hansen, 2009). Here, $\omega=2\pi f$ is the angular frequency and f is the wave frequency. The propagation constant γ is also called the propagation coefficient, which is a complex number and can be expressed as:

$$\gamma = \alpha + jk \tag{6.4}$$

where the propagation wavenumber $k=\omega/c$ is also called the phase coefficient, α is called the attenuation coefficient, and c is the speed of sound in the medium. The units for the real and imaginary parts of the propagation constant γ are $N_p\,m^{-1}$ and $rad\,m^{-1}$, respectively. The attenuation coefficient α describes the reduction in amplitude of a progressive wave with distance in the propagation direction by $e^{-\alpha x}$. This coefficient is actually the amplitude attenuation coefficient instead of the energy attenuation coefficient.

When a sound wave is incident upon a textile layer, some of it will be reflected back to the incidence media and some will be transmitted through the textile layer. The fraction of incident energy that is transmitted is termed the transmission coefficient τ. If there is no other path for energy dissipation or reflections, the relationship between the transmission coefficient and the attenuation coefficient is:

$$\tau = \alpha^2 \tag{6.5}$$

The transmission loss TL is defined as the logarithm of the reciprocal of the transmission coefficient as:

$$TL = \quad 10\log_{10}\tau \tag{6.6}$$

In general, the transmission loss through a porous layer depends upon the angle of incidence and is a function of material density and thickness, flow resistivity, and sound frequency. In the low frequency range (eg, less than 100 Hz), the transmission loss is usually less than 20 dB, but can be greater than 20 dB in the high frequency range (eg, greater than 2000 Hz). The transmission loss usually increases with the material density, thickness, flow resistivity, and frequency.

For a sound absorbing textile panel, the TL value at low frequencies is controlled by the stiffness of the panel (Bies and Hansen, 2009). At the frequency of the first panel resonance, the transmission of sound is high and consequently, the TL passing through the panel is low due to in part the damping in the system. At frequencies above the first panel resonance, a generally broad frequency range is encountered, in which transmission loss is controlled by the surface density of the panel. In this frequency range (mass law range) the transmission loss increases with frequency at the rate of 6 dB per octave. At still higher frequencies in the region of the critical frequency, coincidence is encountered. At very high frequencies, the transmission loss again rises. The rise in this region is of the order of 9 dB per octave.

In practice, it is desirable to characterize the transmission loss of a particular panel with a single number descriptor to facilitate comparison of the performance of different panels. For this reason, a single number rating scheme called STC (or sound transmission class) has been introduced (ASTM E413-10, 2010). To determine the STC for a particular partition, a curve fitting technique is used to fit the measured or calculated one-third octave transmission loss data for the panel. The ISO standard uses a Sound Reduction Index (R_i) instead of the Sound Transmission Loss, and a Weighted Sound Reduction Index (R_w) instead of the Sound Transmission Class (STC). Although different terminology is used, the determination methods are similar (ISO 717-1:2013, 2013).

5.1.3 Absorption

The absorption coefficient at a boundary, also called the absorption factor or sound power absorption coefficient, is defined as the fraction of the incident acoustic power arriving at the boundary that is not reflected, and is therefore regarded as being absorbed by the boundary (Morfey, 2001):

$$\alpha = 1 - \frac{E_\mathrm{r}}{E_\mathrm{i}} \tag{6.7}$$

where E_r and E_i are the reflected and the incident acoustic power arriving at the boundary, respectively. The absorption coefficient, α is a function of frequency and the incident wave direction. The (random incidence) statistical absorption coefficient can be obtained by:

$$\alpha_\mathrm{st} = \frac{1}{\pi}\int_0^{2\pi} \mathrm{d}\varphi \int_0^{\pi/2} \alpha(\theta)\cos\theta \sin\theta \mathrm{d}\theta \tag{6.8}$$

where θ and φ are the angles of elevation and azimuth, respectively. The Sabine absorption coefficient is the random incidence absorption coefficient deduced from the reverberation time measurement via the Sabine equation (Bies and Hansen, 2009). The reverberation time is the time taken for the acoustic energy in an initially steady reverberant sound field to decay by 60 dB (Morfey, 2001).

Porous material absorbers are materials where sound propagation occurs in a network of interconnected pores (open pore structure) in such a way that viscous and thermal effects cause acoustic energy to be dissipated. Air is a viscous fluid, and consequently sound energy is dissipated via friction with the pore walls. As well as viscous effects, there will be losses due to thermal conduction from the air to the absorber material, which might be more significant at low frequencies. For a porous absorber to create significant absorption, it needs to be placed somewhere where the particle velocity is high (Morfey, 2001). The particle velocity close to a room boundary (wall) is usually zero, so the parts of an absorber farthest from a backing surface are often most effective in absorbing sound (Kuttruff, 2009). The thickness of a layer of acoustic absorbent material needs to be greater than 1/10 wavelength of the sound to have significant absorption, and should be about 1/4 wavelength thick to absorb most incident sound.

It is often difficult to obtain absorption at low frequencies with porous textile absorbers because the required thickness of the material is large and the sound absorbing layers are often placed at room boundaries where the absorbers are inefficient due to the low particle velocity. Resonant absorbers might be a solution. There are two common forms of resonant absorbers: membrane/panel absorbers and Helmholtz absorbers. A membrane/panel absorber is a sheet of vinyl or plywood, which is free to vibrate; while for a Helmholtz absorber, the mass is a "plug" of air in the opening of a perforated sheet. The spring in both cases is provided by air enclosed in the cavity. Best performance can be obtained by placing a porous textile absorbent in the neck of the Helmholtz resonator or just behind the membrane in the panel absorber. The resonant frequency of this type of absorber can be tuned to the frequency of interest.

Sound absorbers are usually employed for modifying the sound reverberation in a room, suppressing undesired sound reflections from remote walls (echoes), and reducing the acoustical energy density and, hence, the sound pressure level in noisy rooms. There are two standard methods of measuring absorption coefficients. One is for measuring normal incidence absorption coefficients in an impedance tube and the other is for measuring random incidence absorption coefficients in a reverberation room (Bies and Hansen, 2009).

5.1.4 Scattering

The degree of scattering and absorption in a room is an important factor related to the acoustic properties of a room. Although scattered polar responses can give much information about the scattering from a textile surface, it yields too much detail, and a single value is desired in practice to allow easy comparison of diffuser effectiveness. There is also a need for a scattering coefficient to evaluate the amount of dispersion generated by a surface to allow accurate predictions using geometric room acoustic models (Cox and D'Antonio, 2009). The geometric models use a scattering coefficient to determine the proportion of the reflected energy that is reflected in a specular manner and the proportion that is scattered.

Specular reflection indicates a reflection that obeys the Snell's law, ie, the angle of reflection is equal to the angle of incidence. The scattering coefficient s_θ for an incident wave arriving at the boundary with an angle θ is defined as the value calculated by one minus the ratio of the specularly reflected acoustic energy to the total reflected acoustic energy,

$$s_\theta = 1 - \frac{E_s}{E_r} \tag{6.9}$$

where E_s is the specularly reflected acoustic energy and E_r is the total reflected acoustic energy from the incident acoustic power arriving at the boundary. Theoretically, s_θ takes a value between 0 and 1, where 0 means a totally specularly reflecting surface and 1 means a totally scattering surface (ISO, 17497-1, 2004). The random incidence scattering coefficient s is defined as the value calculated by one minus the ratio of the specularly reflected acoustic energy to the total acoustic energy reflected from a surface in a diffuse sound field.

The directional diffusion coefficient $d_{\theta,\varphi}$ is a measure of the uniformity of diffusion produced by a surface for one sound source (Cox and D'Antonio, 2009),

$$d_{\theta,\varphi} = \frac{\left(\sum_{i=1}^{n} 10^{L_i/10}\right)^2 - \sum_{i=1}^{n} (10^{L_i/10})^2}{(n-1)\sum_{i=1}^{n} (10^{L_i/10})^2} \quad (6.10)$$

where the subscript θ is the angle of incidence relative to the reference normal of the surface, and φ indicates the azimuth angle, n is number of receivers, L_i are a set of sound pressure levels in a polar response of the scattering sound. The directional diffusion coefficient has a value between 0 and 1, corresponding to that, one receiver receives nonzero scattered sound pressure or complete diffusion, respectively. The random incidence diffusion coefficient, d, calculated from weighting the directional diffusion coefficients for the difference source positions, is a measure of the uniformity of diffusion for a representative sample of sources over a complete semicircle for a single plane diffuser, or a complete hemisphere for a hemispherical diffuser. The normalized directional diffusion coefficient, $d_{\theta,\varphi,n}$, is the directional diffusion coefficient of the test specimen normalized to that of a reference flat surface. The normalized diffusion coefficient, d_n, is the random incidence diffusion coefficient determined from the normalized directional diffusion coefficient.

Both the scattering and diffusion coefficients are simplified representations of the true reflection behaviour (Cox and D'Antonio, 2009). The purpose of the diffusion coefficient is to enable the design of diffusers and to allow acousticians to compare the performance of surfaces for room design and performance specifications, while the scattering coefficient is to characterize surface scattering for use in geometrical room modelling programs. The diffusion coefficient is different from, but related to, the random incidence scattering coefficient. While the scattering coefficient is a rough measure that describes the degree of scattered sound, the diffusion coefficient describes the directional uniformity of the scattering, ie, the quality of the diffusing surface.

The difference between diffusion and scattering coefficients is the emphasis on which information is most important to be preserved in the data reduction. For diffuser designers, it is the uniformity of all reflected energy which is most important; for room acoustic modeling, it is the amount of energy scattered away from specular angles. The difference between the definitions may appear subtle, but it is significant. The scattering coefficient method gives a quick and rough estimate of the scattering process, and it should not be used to evaluate the worth of surfaces when designing or specifying diffusers (Cox and D'Antonio, 2009). The scattering coefficient is only concerned with how much energy is moved from the specular direction, it does not measure the quality of dispersion. For this reason, diffusing surfaces need to be evaluated using the diffusion coefficient when the quality of scattering is important. The diffusion coefficient should not be used in geometric room acoustic models as its definition is not compatible with the surface scattering models used in current geometric algorithms.

5.2 Flow resistance measurement

Methods for measuring the (static) flow resistivity can be categorized by the direct airflow method (ISO 9053, 1991; ASTM C522-03, 2009), the alternating airflow method (ISO 9053, 1991; Dragonetti et al., 2011), the comparative method (Stinson and Daigle, 1988), and various acoustic methods (Ren and Jacobsen, 1993; Woodcock and Hodgson, 1992; Picard et al., 1998; Doutres et al., 2010; Tao et al., 2015). The direct airflow method measures the pressure drop across a test specimen and the flow velocity through the test specimen separately in a steady air flow, and then calculates the ratio between them. The alternating airflow method only measures the pressure difference across the test specimen or the sound pressure ratio between two coupled cavities, but the flow velocity generated by an oscillating piston is assumed to be known or the volume velocities in both cavities are the same. The comparative method places the test specimen in series with a material with a calibrated flow resistivity and computes the flow resistivity according to the ratio of pressure drops across each element. The acoustic methods are normally carried out in an impedance tube with a loudspeaker and several microphones at low frequencies.

5.2.1 The direct airflow method

The flow resistance of a sample of porous material may be measured using an apparatus that meets the requirements of ISO 9053-1991 or ASTM C522-03, such as that illustrated in Fig. 5.1. Small flow rates from 5×10^{-4} to 5×10^{-2} m/s are recommended to avoid introduction of undamping effects and nonlinear relationships between the pressure drop and the velocity (Bies and Hansen, 2009; Cox and D'Antonio, 2009).

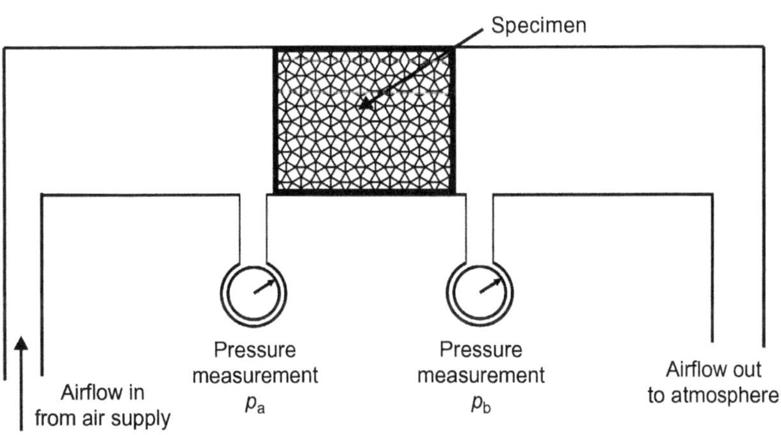

Fig. 5.1 The basic principle for the direct airflow method.

The flow resistivity of the test specimen shown in Fig. 5.1 is calculated from the measured quantities as follows:

$$\sigma = \frac{\Delta p\, S}{V_0\, d} \tag{6.11}$$

where $\Delta p = p_a - p_b$ is the air pressure difference in Pascals across the test specimen, S is the cross section area in square meters of the test specimen, V_0 is the volumetric airflow rate in cubic meters per second passing through the test specimen and d is the thickness in meters of the test specimen in the direction of flow. The measurement cell can be rectangular or cylindrical shaped. If a cylindrical duct is used, its inner diameter should be greater than 95 mm. Square ducts are preferred over rectangular ducts, and the length of the sides should be no less than 90 mm. The length of the measurement duct should be 100 mm greater than the thickness of the test specimen. The lowest flow speed from the steady airflow should be 5×10^{-4} m/s. More details about the equipment, procedure, precision, and report format of the test can be found in the standards (ISO 9053, 1991; ASTM C522-03, 2009).

5.2.2 *The alternating airflow method*

In the alternating airflow method, a piston is used to generate a low frequency alternating airflow through the test specimen and a condenser microphone is used to obtain the root mean square (rms) value of the pressure change in front of the porous material. Fig. 5.2 shows the basic principle of the experiment setup.

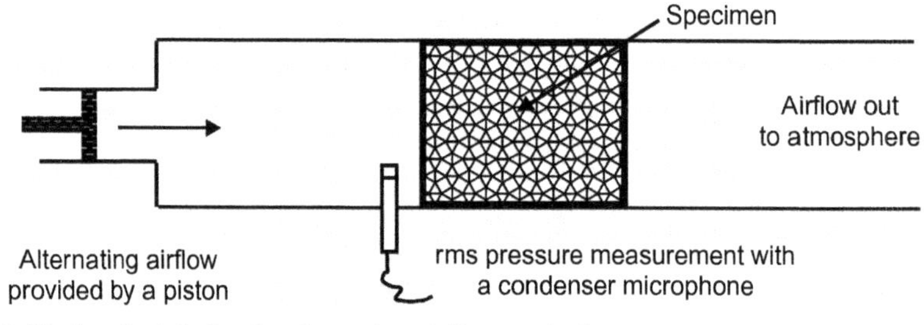

Fig. 5.2 Basic principle for the alternating airflow method.

The frequency of the alternating airflow generated by the piston displacement is recommended to be around 2 Hz, and the rms value of the airflow velocity is recommend to be between 0.5 and 4 mm/s, which can be calculated by

$$u_{rms} = \frac{\pi f h A_p}{\sqrt{2}\, S} \tag{6.12}$$

where f is the frequency of the alternating airflow generated by the piston, h is the peak-to-peak amplitude of the piston displacement, A_p is the cross section area of the piston, and S is the cross section area of the test specimen. By using the relationship between the pressure variation and volume variation (Bjor, 2009), the flow resistivity of the test specimen can be calculated using Eq. (6.13):

$$\sigma = \frac{S\,p}{2\pi\, d\, f\, \frac{A_p\, h}{2\sqrt{2}} \sqrt{1 - \left(\frac{V\,p}{P_a\, \gamma\, \frac{A_p\, h}{2\sqrt{2}}} \right)^2}} \tag{6.13}$$

where p is the sound pressure in Pascals measured by the microphone, P_a is the atmospheric air pressure, d is the thickness of the test specimen in the direction of flow, V is the volume of the vessel between the piston and the test specimen, and $\gamma = 1.4$ is the ratio of specific heat for air. When the sound pressure level is less than 140 dB, the preceding equation can be approximated by

$$\sigma = \frac{\sqrt{2}\, p_{\mathrm{ref}}\, S}{\pi\, d\, f\, h\, A_p} 10^{\frac{L}{20}} \tag{6.14}$$

where $p_{\mathrm{ref}} = 20\ \mu\mathrm{Pa}$ is the reference sound pressure level and L is the measured sound pressure level in dB, so the flow resistivity can be easily measured with a piston and a sound level meter. For typical instrumentation, the maximum diameter of test pieces is about 100 mm, the diameter of the piston is about 10 mm with a stroke length of about 5 mm, and the specific flow resistance measurement range is approximately from 10 to $10^5\,\mathrm{N\,s\,m^{-3}}$. Reproducibility of such measurements might be a problem because an error of about 2.5% is found with repeat measurements on one sample, and about 5% with five samples cut from the same material in the same laboratory, and about 15% between laboratories (Garai and Pompoli, 2003).

5.2.3 Acoustic measurement methods

There are many acoustical methods proposed for measuring flow resistivity (Delany and Bazley, 1971; Smith and Parott, 1983). A method that uses a standard impedance tube directly to measure the static flow resistivity without any additional requirements to tube modification or sensor location change is described in ISO Standard, 10534-2 (1998) and by Tao et al. (2015). In the method, the specific acoustic impedance on the front surface of the test specimen is measured first by using the traditional transfer function method with the test specimen being placed against and with a known interval to the rigid termination, and then the characteristic impedance, the propagation constant, and the static flow resistivity are calculated based on the obtained impedance transfer functions.

Fig. 5.3 shows a diagram of the impedance tube designed for the ISO 10534.2 implementation, where the test specimen thickness is $2l$ and the back cavity depth (the distance between the back surface of the specimen and the rigid termination) is L. The specific acoustic impedance at the front surface of the test specimen can be determined by (ISO Standard, 10534-2, 1998):

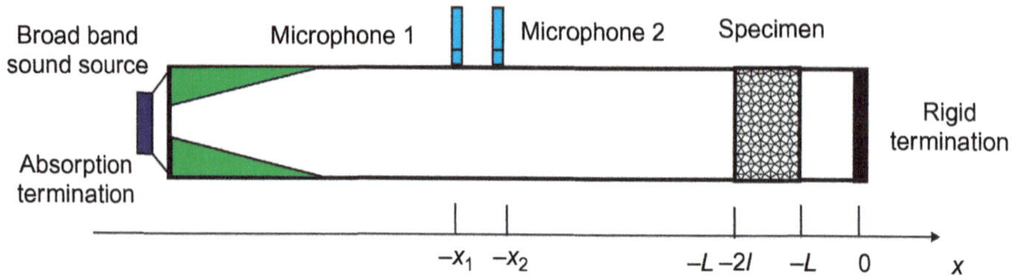

Fig. 5.3 A diagram of an impedance tube for the ISO 10534.2 method.

$$Z_s = \rho_0 c_0 \frac{1+R}{1-R} \tag{6.15}$$

where ρ_0 is the air density, c_0 is the speed of sound, and R is the reflection coefficient, which can be obtained using ISO Standard 10534-2 (1998):

$$R = \frac{H_{12} - H_i}{H_r - H_{12}} e^{2jk_0 x_1} \tag{6.16}$$

In the equation, $H_{12}=p_2/p_1$, p_1 and p_2 are the measured sound pressures at microphones 1 and 2, respectively. $H_i = e^{-jk_0(x_1-x_2)}$, $H_r = e^{jk_0(x_1-x_2)}$, x_1 and x_2 are the location coordinates of Microphone 1 and 2, respectively, and k_0 is the wave number. In general, the wave number k_0 depends on the propagation attenuation in the tube, but can be approximated to $2\pi f/c_0$ for convenience, where f is the frequency.

On the other hand, the specific acoustic impedance at the test specimen's front surface can be formulated with the impedance transfer function (Kinsler et al., 2000):

$$Z_s = Z_m \frac{Z_L + jZ_m \tan(2k_m l)}{Z_m + jZ_L \tan(2k_m l)} \tag{6.17}$$

where Z_m, k_m and Z_L are respectively the characteristic impedance, propagation constant, and the acoustic impedance at the back surface of the test specimen. For a rigid termination:

$$Z_L = -j\rho_0 c_0 \cot(k_0 L) \tag{6.18}$$

With a test specimen placed against a rigid termination, $L=0$ and $Z_L \to -\infty$, so Eq. (6.17) becomes:

$$Z_m = jZ_s \tan(2k_m l) \tag{6.19}$$

Assuming that the specific acoustic impedance on the front surface of the test specimen when L is zero and nonzero is measured as $Z_{s,0}$ and $Z_{s,L}$ respectively, the propagation constant k_m can be obtained by substituting Eqs. (6.18) and (6.19) into Eq. (6.17) as Eq. (6.20):

$$k_{\mathrm{m}} = \pm\frac{1}{2l}\arctan\sqrt{\frac{Z_{\mathrm{L}}}{Z_{\mathrm{s},0}} - \frac{Z_{\mathrm{s},L}\left(Z_{\mathrm{s},0}+Z_{\mathrm{L}}\right)}{Z_{\mathrm{s},0}^{2}}} \tag{6.20}$$

By substituting the calculated k_{m} into Eq. (6.19), the characteristic impedance of the test specimen Z_{m} can be obtained. For porous materials, the static flow resistivity can be calculated by Eq. (6.21) (Doutres et al., 2010):

$$\sigma = -\lim_{\omega\to 0}\left[\mathrm{Im}\left(Z_{\mathrm{m}}k_{\mathrm{m}}\right)\right] \tag{6.21}$$

When the test specimen thickness is rather small compared to the wavelength, $\tan(2k_{\mathrm{m}}l) \approx 2k_{\mathrm{m}}l$ in Eq. (6.19), then Eq. (6.21) can be simplified to Eq. (6.22):

$$\sigma = -\lim_{\omega\to 0}\left\{\mathrm{Im}\left[\frac{Z_{\mathrm{L}} - Z_{\mathrm{s},L}\left(Z_{\mathrm{L}}+Z_{\mathrm{s},0}\right)/Z_{\mathrm{s},0}}{2l}\right]\right\} \tag{6.22}$$

In practical applications, the lower limit of the measurement frequency in ISO 10534.2 is 63 Hz and it is impossible to generate incident sound waves at zero Hz as required in Eq. (6.22). Fortunately, the flow resistivity is not sensitive to frequency when the measurement frequency is sufficiently low (Ren and Jacobsen, 1993), so the determined value of static flow resistivity is expected to be acceptable when the measurement frequency is around a few hundred Hz.

5.3 Transmission loss measurement

The transmission coefficient is the fraction of incident energy transmitted through a partition or a textile layer when a sound wave is incident upon it, and the transmission loss is the logarithm of the reciprocal of the transmission coefficient. Two types of transmission loss are commonly measured with a layer of textile material. The random incidence transmission loss is measured in a reverberation room using the reverberant room method, while the normal incidence sound transmission loss is measured in an impedance tube using the transfer matrix method.

5.3.1 *The reverberant room method*

In general, the transmission loss depends upon the angle of incidence of the incident sound (Bies and Hansen, 2009). Random incidence transmission loss refers to the response of test specimens exposed to a diffuse incident sound field in testing laboratories. The test results are therefore most directly relevant to the performance of similar specimens exposed to similar sound fields. The random incidence transmission loss of a partition is measured in a laboratory by placing the partition in an opening between two adjacent reverberant rooms designed for such tests as shown in Fig. 5.4 (ASTM E90-09, 2009; ISO, 10140-2, 2010).

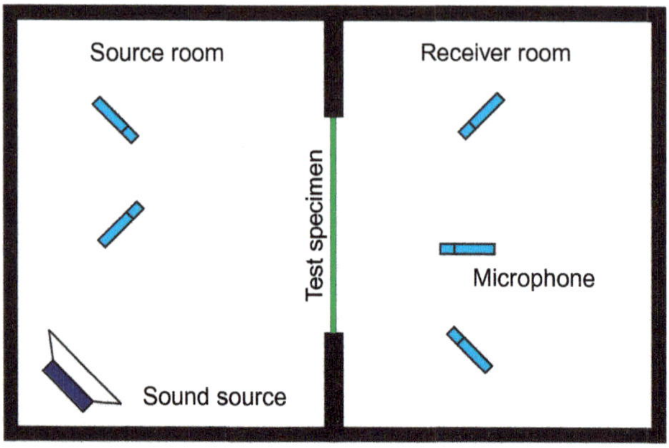

Fig. 5.4 Partition transmission loss measurements in a laboratory with two adjacent reverberant rooms.

Sound is introduced into one of the rooms, referred to as the source room, and part of the sound energy is transmitted through the test partition into the second room, referred to as the receiver room. The resulting mean space average sound pressure levels in the source and receiver rooms are measured and the difference in levels, called the noise reduction, *NR*, is determined. The random incidence transmission loss can be obtained by:

$$TL = NR + 10\log_{10}\left(\frac{A}{S\bar{\alpha}}\right) \tag{6.23}$$

where A is the panel area, S is the receiver room total surface area, including that of the test partition, and $\bar{\alpha}$ is the mean Sabine absorption coefficient in the receiver room, including loss back through the test partition.

The ISO 10140 series standard concerns laboratory measurement of the sound insulation of building elements (ISO 10140-1:2010, 2010; ISO 10140-2:2010, 2010; ISO 10140-3:2010, 2010; ISO 10140-4:2010, 2010; ISO 10140-5:2010, 2010). It consists of five parts under the general title of "Acoustics–Laboratory measurement of sound insulation of building elements", and the subtitles are: application rules for specific products, measurement of airborne sound insulation, measurement of impact sound insulation, measurement procedures and requirements for test facilities and equipment. ISO 10140-1 indicates the appropriate test procedure for elements and products. For certain types of elements/products, the standard contains additional and more specific instructions about quantities and test element sizes and about preparation, mounting, and operating conditions. ISO 10140-2 gives a complete procedure for sound insulation measurements according to ISO 10140-4 and ISO 10140-5. The size of the partition required for general tests is about $10\,m^2$.

When conducting a transmission loss test, care must be taken to ensure that all other acoustic transmission paths are negligible. For example, any flanking paths must contribute an insignificant amount to the total energy transmitted. In laboratories designed to satisfy the requirements of this test method, the intent is that the only significant path for sound transmission between the rooms is through the test specimen. This is not generally the case in buildings where there are often many other paths for sounds. Consequently, sound ratings obtained using this test method do not relate directly to sound isolation in buildings. They represent an upper limit to what would be measured in a field test. Field tests should be performed according to other standards, such as ASTM E336-14 (2014).

The reverberant room test method covers the laboratory measurement of sound transmission loss of building partitions such as walls of all kinds, operable partitions, floor-ceiling assemblies, doors, windows, roofs, panels, and other space-dividing elements. Testing laboratories are designed so the test specimen constitutes the primary sound transmission path between the two test rooms and so approximately diffuse sound fields exist in the rooms. The sound transmission loss of a partition may also be determined using a single reverberant room as the source room and a not too reverberant space as the receiver room. In this case, the power incident on the partition and the transmitted power may be determined by measuring the average of the active sound intensity very close (500–100 mm) to the panel on the receiver room side (ISO, 15186-1, 2000; ISO, 15186-2, 2003; ISO, 15186-3, 2002).

5.3.2 The transfer matrix method

Sometimes, it is necessary to characterize the sound insulation characteristics of materials using a less expensive and less time consuming approach than the reverberant room method, or to have small samples tested when larger samples are impossible to construct or to transport to a laboratory. In these cases, the transfer matrix method can be used to measure the normal incidence sound transmission of the materials or samples. The transfer matrix can be used to determine additional acoustic properties of the material and to allow calculation of acoustic properties of built-up or composite materials by the combination of their individual transfer matrices (ASTM E2611-09, 2009).

Transmission loss is strongly dependent on boundary conditions inherent in the method and details of the way that the material is mounted. This must be considered in the interpretation of the results obtained by this test method. The application of materials into acoustical system elements will probably not be similar to this test method and therefore results obtained by this method may not correlate with performance in practical situations. The quantities are measured as a function of frequency with a resolution determined by the sampling rate, transform size, and other parameters of a digital frequency analysis system. The usable frequency range depends on the diameter of the tube and the spacing between the microphone positions. An extended frequency range may be obtained by using tubes with various diameters and microphone spacing.

A simple schematic of an impedance tube is shown in Fig. 5.5. It essentially consists of a straight tube with one end connected to a sound source and the other end with the capability to hold a material sample whose properties are to be measured. A pair of microphones separated by a specified distance is connected to this tube with the help of microphone holders. These microphones are connected to a digital signal analyzer via signal conditioners (preamplifiers) and a data acquisition system. A function generator with an equalizer and an amplifier is used to power the sound source in the impedance tube. Unlike absorption coefficient measurements where a rigid backing is used, a hollow tube of the same diameter as the upstream tube with a pair of microphone holders is used on the downstream side of the test specimen for transmission loss measurement. Two different termination conditions (anechoic and rigid backing) are used during the transmission loss measurements.

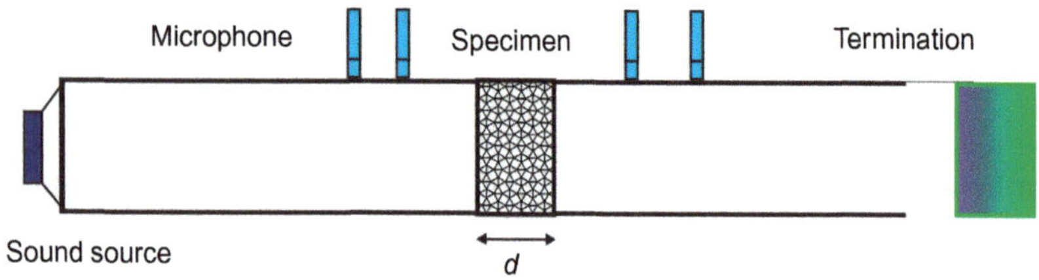

Fig. 5.5 Setup for normal incidence sound transmission loss measurements with an impedance tube.

Four microphones, at two locations on each side of the test specimen, are mounted so that the diaphragms are flush with the interior surface of the tube. Plane waves are generated in the tube using a broadband signal from a sound source. The resulting standing wave pattern is decomposed into forward and backward traveling components by measuring the sound pressure simultaneously at the four locations and examining their relative amplitude and phase. The acoustic transfer matrix,

$$\begin{bmatrix} p \\ v \end{bmatrix}_{x=0} = \begin{bmatrix} T_{11} & T_{12} \\ T_{21} & T_{22} \end{bmatrix} \begin{bmatrix} p \\ v \end{bmatrix}_{x=d} \tag{6.24}$$

is calculated from the pressure p and particle velocity v on either side of the test specimen. This transfer matrix relates the sound pressures and normal acoustic particle velocities on the two faces of the test specimen and completely describes the plane wave characteristics of the test specimen (ASTM E2611-09, 2009). To obtain the transfer matrix, it is necessary to perform tests using two different terminations. Generally, nearly-anechoic and open terminations are used. After the elements of the matrix are obtained, the normal incidence transmission loss can be calculated using the transfer matrix elements as in Eq. (6.25) (ASTM E2611-09, 2009):

$$TL = 20\log_{10}\left|\frac{T_{11} + T_{12} / \rho_0 c_0 + \rho_0 c_0 T_{21} + T_{22}}{2e^{jkd}}\right| \tag{6.25}$$

where d is the thickness of the test specimen, ρ_0 is the air density, c_0 is the speed of sound, the wavenumber $k = 2\pi f/c_0$, and f is the frequency.

This measurement of the normal incidence sound transmission loss standardized in ASTM E2611-09 uses a plane wave tube instrumented with four microphones and a termination consisting of an adaptable acoustic load (four-microphone-two-load). The two loads are typically a minimally reflecting termination (eg, anechoic termination), and a termination reflecting a portion of incident wave (eg, open termination). A minimum of six transfer functions need to be measured. A different method has been proposed, which uses three microphones in a slightly modified classical impedance tube, where the third microphone is flush mounted on the movable hard termination. The proposed method requires fewer transfer function measurements (four in the general case) than the standard method introduced in this section (Salissou et al., 2012).

5.4 Absorption coefficient measurement

Two techniques commonly used to perform absorption coefficient measurements are the impedance tube method and the reverberant room method. The impedance tube method is for measuring normal incidence absorption coefficients while the reverberant room method is for measuring random incidence absorption coefficients. Usually, a material with a large normal incidence absorption coefficient also has a large random incidence absorption coefficient; however, there is no clear definitive relationship between the two absorption coefficients (McGrory et al., 2012). There are also other methods for measuring the absorption coefficients of material under different situations (Cox and D'Antonio, 2009). For example, the two microphone techniques can be used to obtain oblique incidence absorption coefficients in a free field, based on the measured transfer functions between the two microphone positions and time gating (windowing) techniques that can be applied for in situ measurements (ISO, 13472-2, 2010).

5.4.1 The impedance tube method

For the impedance tube measurements shown in Fig. 5.6, a rigid wall tube/duct is used with a loudspeaker at one end and the test specimen terminates the other end of the tube (ISO Standard, 10534-2, 1998; ASTM E2611-09, 2009; ASTM E1050-10, 2010). The test specimen must be arranged in the same way as it is used in practice such as the distance in front of a rigid wall. An absorbing termination is placed in front of the loudspeaker to avoid the reflection from the source end. A movable probe microphone, which must be small enough not to distort the sound field, is used to measure the sound pressure maxima and minima of the standing wave in the tube at a single frequency. The tube should be long enough to permit the formation of at least one maximum and one minimum of the pressure distribution at the lowest frequency of interest. Its lateral dimensions should be smaller than a certain fraction of

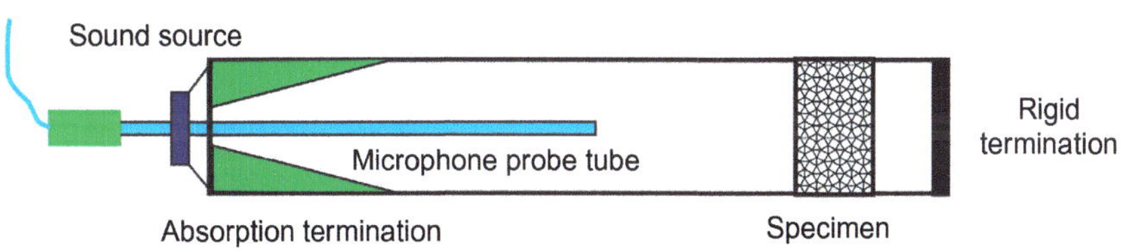

Fig. 5.6 Setup for normal incidence absorption coefficient measurements within an impedance tube based on the standing wave method.

the wavelength at the highest frequency to avoid high order modes which do not have constant lateral pressure distributions. On the other hand, the cross section of the tube must not be too small to avoid large wave attenuation due to losses at the wall surface. Generally at least two tubes of different dimensions are needed in order to cover the frequency range from about 100 Hz to 5000 Hz.

Assuming the normal plane wave pressure reflection coefficient at the end of duct with the test specimen is R, the steady state sound pressure at location x and at time t in the tube is given by

$$p(x,t) = p_{\mathrm{i}} + p_{\mathrm{r}} = p_{\mathrm{A}}\mathrm{e}^{\mathrm{j}(\omega t - kx)} + Rp_{\mathrm{A}}\mathrm{e}^{\mathrm{j}(\omega t + kx)} \tag{6.26}$$

where p_A is the magnitude of the normal incidence plain wave, c_0 is the speed of sound, the angular frequency $\omega = 2\pi f$, the wavenumber $k = \omega / c_0$, and f is the frequency. Its magnitude is

$$|p(x,t)| = p_{\mathrm{A}}\sqrt{(1+|R|)^2 + 2|R|\cos(2kx+\Psi)} \tag{6.27}$$

where Ψ relates to the phase change caused by the reflection. When $2kx + \Psi = 2n\pi$ or $2(n+1)\pi$, the maximum and minimum values of the sound pressure can be obtained as:

$$p_{\max} = p_{\mathrm{A}}(1+|R|) \tag{6.28}$$

$$p_{\min} = p_{\mathrm{A}}(1-|R|) \tag{6.29}$$

so the reflection coefficient can be derived:

$$|R| = \frac{p_{\max} - p_{\min}}{p_{\max} + p_{\min}} \tag{6.30}$$

and the normal incidence absorption coefficient is:

$$\alpha = 1 - |R|^2 = \frac{4p_{\max}p_{\min}}{(p_{\max} + p_{\min})^2} \tag{6.31}$$

If the sound pressure level difference between the maximum and minimum sound pressure is used, the normal incidence absorption coefficient can be obtained by:

$$\alpha = \frac{4 \times 10^{\Delta L/20}}{\left(1 + 10^{\Delta L/20}\right)^2} \tag{6.32}$$

where $\Delta L = 20\log_{10}(p_{\max}/p_{\min})$.

The advantage of the standing wave method is simple, robust, and reliable. Unfortunately, it only measures one frequency at a time, and the procedure for locating minima in the standing wave, which is needed to obtain the phase information, is rather slow, so measuring a large number of frequencies is tedious. To increase the speed of the measurement, the cross-spectrum methods can be used to obtain the normal incidence absorption coefficient of the test specimen in wideband. Fig. 5.7 shows the setup for normal incidence absorption coefficient measurements within an impedance tube by using the cross-spectrum method, where two microphones are used to obtain the spectrum at two locations in the tube (ISO Standard, 10534-2, 1998; ASTM E2611-09, 2009; ASTM E1050-10, 2010).

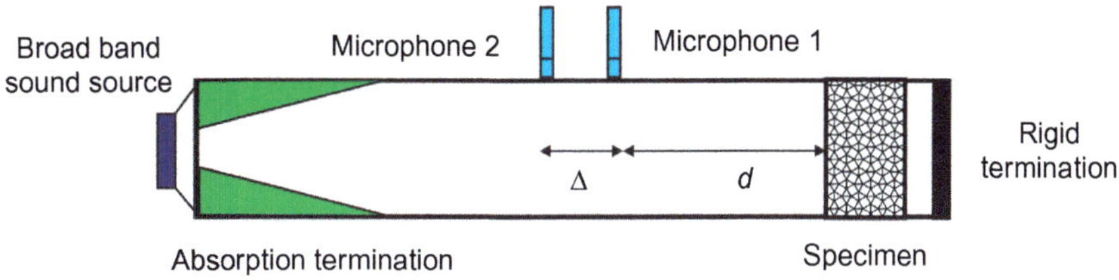

Fig. 5.7 Setup for normal incidence absorption coefficient measurements with an impedance tube based on the cross-spectrum method.

The sound pressure spectrum at microphones 1 and 2 in Fig. 5.7 can be expressed by:

$$S_1(f) = S(f)\left[e^{-jkd} + R(f)e^{jkd}\right] \tag{6.33}$$

$$S_2(f) = S(f)\left[e^{-jk(d+\Delta)} + R(f)e^{jk(d+\Delta)}\right] \tag{6.34}$$

where d is the distance from microphone 1 to the test specimen, Δ is the distance between two microphones, and $S(f)$ is the spectrum of the incidence wave from the broadband sound source. The wavenumber $k=2\pi f/c_0$, f is the frequency, and c_0 is the speed of sound. From Eqs. (6.33) and (6.34), the reflection coefficient spectrum can be obtained as

$$R(f) = e^{-2jkd}\frac{S_2(f) - S_1(f)e^{-jk\Delta}}{S_1(f)e^{jk\Delta} - S_2(f)} \tag{6.35}$$

Once the cross spectrum $H_{12}(f)=S_2(f)/S_1(f)$ between two microphones is obtained, the reflection coefficient spectrum can be written as

$$R(f) = e^{-2jkd}\frac{H_{12}(f) - e^{-jk\Delta}}{e^{jk\Delta} - H_{12}(f)} \tag{6.36}$$

and the normal incidence absorption coefficient is

$$\alpha = 1 - |R|^2 \tag{6.37}$$

The main advantage of using the cross-spectrum method is that it obtains the absorption coefficient of the surface for all frequencies (within limits) at the same time and is much more efficient than the standing wave method. There are restrictions on the microphone spacing (Cox and D'Antonio, 2009). If the microphones are too close together, the transfer function measured tends to be inaccurate because the change in pressure is too small to be accurately measured. If the microphone spacing is too wide, the simultaneous equations become impossible to resolve as the pressure measured at both microphones might be nearly identical at some frequencies.

5.4.2 The reverberation room method

The impedance tube measurement obtains the normal incidence absorption coefficient of a layer of textiles with a small diameter (usually less than 10cm). The results can be used to compare the basic absorption performance of a material and for acoustics simulations. But in practice, the absorbers can be quite large and their structure and configuration may be complex and part of acoustic designs. Furthermore, they will be exposed to real sound fields where the incident sound may come from many directions. The reverberation room method is superior to the impedance tube method in the following aspects. First, the measurement is performed with a diffuse sound field, ie, under conditions which are closer to many practical installations. Second, there are no limitations concerning the type and construction of the absorber. It is well suited for measuring the absorption coefficient of almost any type of wall linings and of ceilings, as well as single seats or blocks of seats, unoccupied or occupied.

Reverberation room method measurements must be carried out in a small room with a volume of at least 100 m^3, preferably between 200 and 300 m^3. The walls of the room should be as smooth and rigid as possible to reduce the reflection losses (ASTM C423-09, 2009; ISO 354, 2003). The room surfaces (including the ceiling) are often arranged to be nonparallel to prevent the formation of standing waves. Additional acoustic diffusers are often used to create more reflecting surfaces and further encourage even distribution of any particular sound field.

The absorption coefficient can be determined by reverberation time measurements in an empty reverberation room and when a certain amount of the material is brought into the reverberation room and is mounted in the same way as it would be in practice. According to the Sabine reverberation time equation (Bies and Hansen, 2009; Cox and D'Antonio, 2009), the reverberation time of the empty reverberation room can be expressed by:

$$T_0 = \frac{55.3V}{c_0 S \alpha_0} \tag{6.38}$$

where S is the surface area of the room, V is the room volume, c_0 is the speed of sound in air, and α_0 is the average absorption coefficient of the empty room, which is often quite small. The Sabine reverberation time equation in the room containing a test specimen is:

$$T_a = \frac{55.3V}{c_0\left[(S - S_a)\alpha_0 + S_a\alpha_a\right]} \tag{6.39}$$

where S_a is the surface area of the test specimen and α_a is the average absorption coefficient of the reverberation room with the test specimen. From the preceding two equations, the average absorption coefficient of the test specimen can be obtained as:

$$\alpha_a = \frac{55.3V}{c_0 S_a}\left[\frac{1}{T_a} - (S - S_a)\frac{1}{ST_0}\right] \tag{6.40}$$

By assuming $S \gg S_a$,

$$\alpha_a \approx \frac{55.3V}{c_0 S_a}\left(\frac{1}{T_a} - \frac{1}{T_0}\right) \tag{6.41}$$

In the measurements, the surface area of the test material should be large enough to measurably affect the reverberation time, but not so large as to seriously affect the diffusivity of the sound field. The ASTM and ISO standards recommend that the area should be between 10 and 12 m^2 with a length-to-breadth ratio between 0.7 and 1.0. In many cases, sound absorption in a reverberation room is affected by things other than the room walls, such as loudspeakers, at low frequencies; stationary and rotating diffuser surfaces, at low and mid frequencies; and air absorption, at high frequencies. All these need to be considered in formal tests.

The value of the measured Sabine absorption coefficient of a textile might be different with different sample sizes, sample distributions and properties of the room in which it is measured. Therefore, when values measured in a laboratory are used to calculate reverberation times and reverberant sound pressure levels in practical auditoria and factories, the calculated values are approximate only.

The reverberation time of a room can be measured using two methods: the decay curve method and the backward integration method (ASTM C423-09, 2009; ASTM E2235-04, 2012; ISO 3382-2, 2008). The decay curve method drives a speaker by a random noise generator in series with a bandpass filter. When the sound is turned off, the rate of the decay of sound in the room is measured simply by a sound pressure level meter attached to a level recorder. The slope, generally measured as the best straight line fit to the recorded decay curve between 5 and 35 dB down from the initial steady state level, is used to determine the decay rate. The reverberation time is two times the decay time from −5 to −35 dB estimated from the line. The backward integration method has greater accuracy, where the reverse time integration transforms the normal room impulse response into a decay plot so that it is possible to evaluate the decay time and consequently the reverberation time (Schroeder, 1965; Morgan, 1997).

5.5 Scattering property measurement

The acoustic quality in a room can be affected by the scattering of sound by the walls and objects inside. The scattering coefficient and the diffusion coefficient are two common measures to describe the scattering produced by a surface. The random incidence scattering coefficient of surfaces caused by surface roughness is measured in a reverberation room, while the directional diffusion coefficient of surfaces is measured in a free field. The scattering coefficient is used to characterize surface scattering in geometrical room modeling, while the diffusion coefficients enable diffuser designers and acousticians to compare the performance of different surfaces.

5.5.1 Measurement of the random incidence scattering coefficient in a reverberation room

The first part of ISO 17497 specifies a method of measuring the random incidence scattering coefficient of surfaces caused by surface roughness (ISO, 17497-1, 2004). The measurements are made in a reverberation room, either in full scale or on a scale model. The measurement results can be used to describe how much the sound reflection from a surface deviates from a specular reflection. The results obtained can be used for comparison purposes and for design calculations with respect to room acoustics and noise control. The method is not intended for characterizing the spatial uniformity of the scattering from a surface.

The scattering coefficient can be calculated by one minus the ratio of the specularly reflected acoustic energy to the total reflected acoustic energy:

$$s_\theta = 1 - \frac{E_\mathrm{s}}{E_\mathrm{r}} \tag{6.42}$$

where the total reflected acoustic energy can be expressed from Eq. (6.7) as:

$$E_\mathrm{r} = (1 - \alpha_\mathrm{a}) E_\mathrm{i} \tag{6.43}$$

where E_i is the total incident acoustic power and α_a is the absorption coefficient. The specularly reflected acoustic energy can be expressed similarly as:

$$E_\mathrm{s} = (1 - \alpha_\mathrm{s}) E_\mathrm{i} \tag{6.44}$$

where α_s is the specular absorption coefficient. By substituting the preceding two equations into Eq. (6.42), the scattering coefficient can be obtained:

$$s_\theta = \frac{\alpha_\mathrm{s} - \alpha_\mathrm{a}}{1 - \alpha_\mathrm{a}} \tag{6.45}$$

The absorption coefficient and the specular absorption coefficient can be measured in a reverberation room.

Fig. 5.8 shows the setup for measuring the random incidence scattering coefficient in a reverberation room. The random incidence absorption coefficient of the test specimen can be obtained using the method mentioned in Section 5.4.2. The specular absorption coefficient is obtained by rotating the test specimen while phase lock averaging the reflected pulses. A circular test specimen is placed on a turntable. While the turntable is rotated, the impulse response is repeatedly measured. The initial parts of the impulse responses corresponding to the direct and specular components are highly correlated and remain unaltered as the test specimen is rotated, while the later parts in

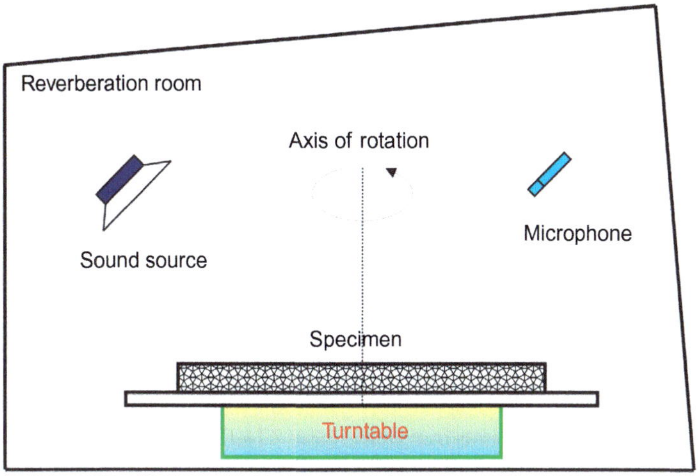

Fig. 5.8 Setup for measuring the random incidence scattering coefficient in a reverberation room.

the impulse responses corresponding to the scattered component are not in phase and depend strongly on the specific orientation. By averaging the reflected pulse pressure while rotating the test specimen, the scattered components are averaged to zero and only the specular energy remains. This impulse response is then backward integrated to give the reverberation time due to the specular reflection component (Schroeder, 1965; Morgan, 1997).

By manipulating these reverberation times, the scattering coefficient can be obtained. In the test, 4 reverberation times are measured, where T_1 is that without the test specimen when the turntable is off, T_2 is that with the test specimen when the turntable is off, T_3 is that without the test specimen when the turntable is on, and T_4 is that with the test specimen when the turntable is on. The random incidence absorption coefficient of the test specimen can be calculated by:

$$\alpha_a \approx \frac{55.3V}{c_0 S_a}\left(\frac{1}{T_2} - \frac{1}{T_1}\right) \tag{6.46}$$

and the random incidence specular absorption coefficient can be calculated by:

$$\alpha_s \approx \frac{55.3V}{c_0 S_a}\left(\frac{1}{T_4} - \frac{1}{T_3}\right) \tag{6.47}$$

In the formal measurements, the speed of sound in air and the air absorption at different measurement times need to be taken into consideration. The diameter of the flat turntable is about 3.6 m, so it might be difficult to have a perfectly flat and circular turntable with a low-noise driving motor. The imperfections in the turntable also need to be compensated for by additional measurements. The measurements must use a deterministic source signal such as the maximum length sequences (MLS) signal to allow phase locked pressure averaging. In a full-scale test, the turntable must move slowly, and the signal periodicity must be longer than the reverberation time in the room. Typically 72 measurements are required in a single rotation so obtaining full scale measurements is rather slow and laborious. Furthermore, since the measurement method is intended to measure surface roughness, the results are only reliable when the structural depth of the test specimen is small compared to its size, for example, less than one sixteenth of the diameter of the turntable. More details can be found in the standard (ISO, 17497-1, 2004).

5.5.2 Measurement of the directional diffusion coefficient in a free field

The second part of ISO 17497 specifies a method of measuring the directional diffusion coefficient of surfaces in a free field (ISO, 17497-2, 2012). The diffusion coefficient characterizes the sound reflected from a surface in terms of the uniformity of the reflected polar distribution. The directional diffusion coefficient is a frequency dependent value derived from the polar distribution of the scattered sound (Cox and D'Antonio, 2009). First the scattering from a surface is measured in terms of a polar distribution. Then the diffusion coefficient is evaluated at 1/3 octave bandwidth intervals, which has the advantage of smoothing out some of the local variations in the polar responses.

There are various techniques for measuring a scattered polar distribution. A sound source is used to irradiate the test surface, and measurement microphones at radial positions in front of the surface are used to obtain the pressure impulse response. The microphone positions usually map out a semicircle or hemisphere, for a single plane or hemispherical measurement, respectively. Once the pressure impulse responses are measured with the MLS signal or other signals, time gating is used to separate the reflections from the incident sound. Polar response measurements can be made in a single plane on a semicircle or over a hemisphere depending on the type of diffuser. In the measurements, either a fixed microphone array or a movable single microphone can be used.

Many statistical parameters can be used to describe a diffusion coefficient from measurements of polar sound distributions. In the standard, an autocorrelation function is used to measure the scattered energy's spatial similarity with receiver angles. A surface which scatters sound uniformly to all receivers produces high values in a spatial autocorrelation function; conversely a surface which concentrates scattered energy in one direction has a low value (Cox and D'Antonio, 2009). By using the directional diffusion coefficient calculated by Eq. (6.10), the normalized directional diffusion coefficient can be obtained by:

$$d_{\theta,\varphi,n} = \frac{d_{\theta,\varphi} - d_{\theta,\varphi,r}}{1 - d_{\theta,\varphi,r}} \tag{6.48}$$

where $d_{\theta,\varphi,r}$ is the directional diffusion coefficient of the reference flat surface of the same overall size as the test specimen.

5.6 Summary

The acoustic properties of textiles can be characterized in three categories as propagation, absorption, and scattering, which can be described by flow resistance, transmission loss, absorption coefficient, and scattering coefficient. Flow resistivity is an intrinsic property of a textile and is a measure of how easily air can enter a porous textile material and the resistance that the air flow meets through the material. A direct airflow method, an alternating airflow method, and an acoustic method for measuring the flow resistivity of a textile are described. The fraction of incident energy that is transmitted through a layer of textile is called the transmission coefficient, and the transmission loss is defined as the logarithm of the reciprocal of the transmission coefficient. The random incidence transmission loss of a layer of textile can be measured using the reverberant room method, while the normal incidence sound transmission loss of a layer of textile can be measured in an impedance tube using the transfer matrix method.

The absorption coefficient at a boundary is defined as the fraction of the incident acoustic power arriving at the boundary that is not reflected. Two techniques commonly used are the impedance tube method and the reverberant room method. The impedance tube method is used for measuring normal incidence absorption coefficients while the reverberant room method is used for measuring random incidence absorption coefficients. In the impedance tube measurements, either the standing wave method or the cross-spectrum method can be used. The main advantage of using the cross-spectrum method is that it obtains the absorption coefficient of the surface for all frequencies at the same time and is more efficient than the standing wave method. The reverberation room method is superior to the impedance tube method because the measurement is performed with a diffuse sound field and there are no limitations concerning the type and construction of the absorber. The absorption coefficient is determined by the reverberation time measured in a reverberation room with and without the test specimen by using either the decay curve method or the backward integration method.

The scattering and absorption in a room are important factors related to the acoustic quality of the room. The scattering coefficient for an incident wave arriving at the boundary is defined as the value calculated by one minus the ratio of the specularly reflected acoustic energy to the total reflected acoustic energy. The directional diffusion coefficient is a measure of the uniformity of diffusion produced by a surface for one sound source. The random incidence scattering coefficient of surfaces caused by surface roughness can be measured in a reverberation room, either in full scale or on a scale model, by using a rotating turntable. The directional diffusion coefficient of surfaces can be measured in a free field through measuring the polar distribution of the scattering from a surface with a fixed microphone array or a movable single microphone. Both the scattering and diffusion coefficients are simplified representations of the true reflection behaviour. The purpose of the diffusion coefficient is to enable the design of diffusers and to allow acousticians to compare the performance of surfaces for room design and performance specifications, while the scattering coefficient is used to characterize surface scattering for use in geometrical room modelling programs.

Based on the acoustic properties of the textiles, acoustics designers can use textiles in buildings and office environments to optimize the sound quality. However, the acoustic quality in rooms depends on a number of factors such as the type and location of the sound source, the geometry and acoustic properties of the room surface, and the environmental background noise, etc. It depends not only on the materials to be used, but also on the structures they formed into and on the locations and geometry in which they are installed. Some applications need large sound absorbing materials or structures while other applications require materials or structures that have large scattering but small absorption properties. When there is a need to reduce sound propagation, large sound transmission loss textiles are needed. Until now, there has been no single material or textile that can solve all acoustics problems. When efficient sound absorption is required, there is a need for light, thin and/or transparent materials or textiles with large absorption coefficients, especially at low frequencies. In fact, light and thin, and even transparent textiles, materials, and structures with high sound absorption or with high acoustic scattering or with high transmission loss at low frequencies are some of the challenges for textile designers and manufacturers.

References

ASTM C423-09, 2009. Standard test method for sound absorption and sound absorption coefficients by the reverberation room method. American Society for Testing and Materials.

ASTM C522-03, 2009. Standard test method for airflow resistance of acoustical materials. American Society for Testing and Materials.

ASTM E90-09, 2009. Standard test method for laboratory measurement of airborne sound transmission loss of building partitions and elements. American Society for Testing and Materials.

ASTM E336-14, 2014. Standard test method for measurement of airborne sound attenuation between rooms in buildings. American Society for Testing and Materials.

ASTM E413-10, 2010. Classification for rating sound insulation. American Society for Testing and Materials.

ASTM E1050-10, 2010. Standard test method for impedance and absorption of acoustical materials using a tube, two microphones and digital frequency analysis system. American Society for Testing and Materials.

ASTM E2611-09, 2009. Standard test method for measurement of normal incidence sound transmission of acoustical materials based on the transfer matrix method. American Society for Testing and Materials.

ASTM E2235-04, 2012. Standard test method for determination of decay rates for use in sound insulation test methods. American Society for Testing and Materials.

Bies, D.A., Hansen, C.H., 2009. Engineering Noise Control—Theory and Practice, fourth ed. Spon Press, London and New York.

Bjor, O.H. (2009). Measurement of specific airflow resistance, Nor 1517 Airflow resistance, Norsonic AS.

Cox, T.J., D'Antonio, P., 2009. Acoustic Absorbers and Diffusers: Theory, Design and Application, second ed. Taylor and Francis, London and New York.

Delany, M.E., Bazley, E.N., 1971. Acoustical properties of fibrous absorbent materials. Appl. Acoust. 3, 105–116.

Doutres, O., Salissou, Y., Atalla, N., Panneton, R., 2010. Evaluation of the acoustic and non-acoustic properties of sound absorbing materials using a three-microphone impedance tube. Appl. Acoust. 71 (6), 506–509.

Dragonetti, R., Ianniello, C., Romano, R.A., 2011. Measurement of the resistivity of porous materials with an alternating air-flow method. J. Acoust. Soc. Am. 129 (2), 753–764.

Garai, M., Pompoli, F., 2003. A European inter-laboratory test of airflow resistivity measurements. Acta Acoust. United Acoust. 89, 471–478.

ISO 354:2003, 2003. Acoustics—Measurement of sound absorption in a reverberation room. International Organization for Standardization.

ISO 717-1:2013, 2013. Acoustics—Rating of sound insulation in buildings and of building elements—Part 1: Airborne sound insulation. International Organization for Standardization.

ISO 3382-2, 2008. Acoustics—Measurement of room acoustic parameters—Part 2: Reverberation time in ordinary rooms. International Organization for Standardization.

ISO 9053, 1991. Acoustics—Materials for acoustical applications—Determination of airflow resistance. International Organization for Standardization.

ISO 10140-1, 2010. Acoustics—Laboratory measurement of sound insulation of building elements—Part 1: Application rules for specific products. International Organization for Standardization.

ISO 10140-2, 2010. Acoustics—Laboratory measurement of sound insulation of building elements—Part 2: Measurement of airborne sound insulation. International Organization for Standardization.

ISO 10140-3:2010, 2010. Acoustics—Laboratory measurement of sound insulation of building elements—Part 3: Measurement of impact sound insulation. International Organization for Standardization.

ISO 10140-4:2010, 2010. Acoustics—Laboratory measurement of sound insulation of building elements—Part 4: Measurement procedures and requirements. International Organization for Standardization.

ISO 10140-5:2010, 2010. Acoustics—Laboratory measurement of sound insulation of building elements—Part 5: Requirements for test facilities and equipment. International Organization for Standardization.

ISO 10534-2, 1998. Acoustics—Determination of sound absorption coefficient and impedance in impedance tubes—Transfer function method. International Organization for Standardization.

ISO 13472-2, 2010. Acoustics—Measurement of sound absorption properties of road surfaces in situ - Part 2: Spot method for reflective surfaces. International Organization for Standardization.

ISO 15186-1:2000, 2000. Acoustics—Measurement of sound insulation in buildings and of building elements using sound intensity—Part 1: Laboratory measurements. International Organization for Standardization.

ISO 15186-2:2003, 2003. Acoustics—Measurement of sound insulation in buildings and of building elements using sound intensity—Part 2: Field measurements. International Organization for Standardization.

ISO 15186-3:2002, 2002. Acoustics—Measurement of sound insulation in buildings and of building elements using sound intensity—Part 3: Laboratory measurements at low frequencies. International Organization for Standardization.

ISO 17497-1, 2004. Acoustics — Sound-scattering properties of surfaces — Part 1: Measurement of the random incidence scattering coefficient in a reverberation room.

ISO 17497-2, 2012. Acoustics — Sound-scattering properties of surfaces — Part 2: Measurement of the directional diffusion coefficient in a free field.

Kinsler, L.E., Frey, A.R., Coppens, A.B., Sanders, J.V., 2000. Fundamentals of Acoustics, fourth ed. John Wiley and Sons Inc, New York.

Kuttruff, H., 2009. Room Acoustics, fifth ed. Taylor & Francis, New York.

McGrory, M., Cirac, D.C., Gaussen, O., Cabrera, D., 2012. Sound absorption coefficient measurement: Re-examining the relationship between impedance tube and reverberant room method. In: Proceedings of Acoustics 2012—Fremantle, Australia.

Morfey, C.L., 2001. Dictionary of Acoustics. Academic Press, San Diego, CA.

Morgan, D.R., 1997. A parametric error analysis of the backward integration method for reverberation time estimation. J. Acoust. Soc. Am. 101, 2686–2693.

Picard, M.A., Solana, P., Urchuegía, J.F., 1998. A method of measuring the dynamic flow resistance and the acoustic measurement of the effective static flow resistance in stratified rockwool samples. J. Sound Vib. 216 (3), 495–505.

Ren, M., Jacobsen, F., 1993. A method of measuring the dynamic flow resistance and reactance of porous materials. Appl. Acoust. 39, 256–276.

Salissou, Y., Panneton, R., Doutres, O., 2012. Complement to standard method for measuring normal incidence sound transmission loss with three microphones. J. Acoust. Soc. Am. 131 (3), EL 216.

Schroeder, M.R., 1965. New method of measuring reverberation time. J. Acoust. Soc. Am. 37, 409–412.

Smith, C.D., Parott, T.L., 1983. Comparison of three methods for measuring acoustic properties of bulk materials. J. Acoust. Soc. Am. 74, 1577–1582.

Stinson, M.R., Daigle, G.A., 1988. Electronic system for the measurement of flow resistance. J. Acoust. Soc. Am. 83, 2422–2428.

Tao, J., Wang, P., Qiu, X., Pan, J., 2015. Static flow resistivity measurements based on the ISO 10534.2 standard impedance tube. Build. Environ. http://dx.doi.org/10.1016/j.buildenv.2015.06.001.

Woodcock, R., Hodgson, M., 1992. Acoustic methods for determining the effective flow resistivity of fibrous materials. J. Sound Vib. 153 (1), 186–191.

Medical textiles testing and quality assurance

6

M. Azam Ali, A. Shavandi
University of Otago, Dunedin, New Zealand

6.1 Introduction

The world textile industry is moving rapidly toward the manufacture of high-added value fiber products (Czajka, 2005; Anand, 2001; Rajendran and Anand, 2002a). This is happening in the healthcare and well-being sectors. Medical textiles are known as healthcare textiles. Medical textiles are the products and constructions used for medical and biological applications and are used primarily for first aid, clinical, and hygienic purposes. They consist of all textile materials used in health and hygienic applications, eg, gauze or bandage materials to scaffolds and a large variety of prostheses or implants in both consumer and medical markets (Islam Kiron, 2015).

The textile sector is one of the most rapidly expanding sectors in the technical textile market. Medical textiles are widely used nowadays in medical and healthcare applications such as surgical applications, wound dressing, orthopedic splints, and cardiovascular grafts as medical devices (Walker, 1999; Rajendran and Anand, 2012). The increasing demand for functional textiles for healthcare applications has changed the conventional purpose of clothing as solely for covering and protecting the body. Therefore, there is a growing interest in production of specific textiles for the top niche markets. Fibers or fabrics can be treated to have a specific function, eg, high water or moisture absorbency, thermal comfort, antimicrobial properties, or UV resistance. The growing interest in production of disposable medical textiles for better healthcare and low cost production leads to the development of new materials with improved clinical hygiene. In this way, the demand for more innovative products also leads to the development of traditional or topical wound management products, eg, swabs, gauzes, and bandage. Additionally, the fast improvement in the field of surgery and medical implants leads to the development of new products for new applications, eg, sutures, cardiovascular devices, and so on.

6.2 Types of medical textiles

Medical textiles can be divided into five categories according to their fiber sources (natural or synthetic) and applications as shown in Scheme 6.1.

Implantable materials include molecularly engineered or bioengineered textiles that react with biological media and are aimed at assisting in tissue healing and repairing

Performance Testing of Textiles. http://dx.doi.org/10.1016/B978-0-08-100570-5.00007-4

processes, for example, wound closure, cardiovascular grafts, artificial tendons, ligaments, cartilage, etc. (Anand, 2001; Buschmann et al., 2015).

Various biodegradable and nonbiodegradable materials are currently used for the medical implants. Though more biodegradable materials have already been developed and made available in the market to replace nonbiodegradable textiles, some material properties (eg, durability and mechanical) need to improve to ensure successful use as implantable medical textiles or devices. On the other hand, conventional metallic implants are slowly losing their market-leading role because of their potential drawbacks, such as the risk of infections, mechanical failure, lack of biocompatibility, and the probable need for a second surgery to remove the device after the fracture heals (Anand et al., 2005). Hence, biodegradable polymers with reinforced composite materials are under evaluation for prosthetic implants. These implants have been used in cardiovascular (eg, stent) to bone and skin regeneration, and helping host cells adhere to the implants and repair or reconstruct the injured and damage tissues (Rajendran and Anand, 2002b). Various knitted, woven, and nonwoven textiles, nano-fibrous matrices, and composites may be used in the production of different medical implants for replacement of damaged tissue, including blood vessels, segments arteries, ligaments, and wound closures (eg, sutures). The implants should perform aptly for the purpose. Therefore, implant performance testing methods have to meet certain criteria to satisfy the desired application requirements. The textile implants should demonstrate satisfactory performance, for example, be biocompatible, have certain porosity to provide the required space, and facilitate blood circulation for human tissue growth and regeneration. Furthermore, in some cases they should be biodegradable or bioabsorbable, but not be toxic and contaminated.

6.2.1 Nonimplantable materials

All types of fiber or textile materials have been developed along and/or with incorporating functional materials (eg, hydrogel, bioactive, silver nanoparticles) for external applications. These medical textiles are largely used for external healthcare applications, which are mainly aimed at protecting and enhancing healing through providing protection against infection, for example, gauzes, bandages, pressure garments, etc. (Rajendran and Anand, 2002a, 2012; Walker, 1999; Islam Kiron, 2015; Buschmann et al., 2015). There are different types of surgical wound dressings or devices, depending on their intended applications. For example, primary wound dressings are placed next to a wound surface and are mainly made of cellulose fibers (Meena et al., 2013). Protective and supporting dressings, bandages, and adhesive tapes are also used for wound administration. However, all dressing products should have some required properties. Materials should decrease the surface of adherence, and not have any loose fibers that may be caught in the wound, causing further injury. Therefore, the dressing has to be stable with spatial structure and easily adsorb wound secretions, and moreover, the process of changing the dressing should be painless.

6.2.2 Healthcare/hygiene products

Healthcare/hygiene products are usually available over the counter and normally used for hygienic purposes to prevent infection and transmission of diseases, provide hygiene, and enhance care in the hospital ward and operating room (Rajendran and Anand, 2012). It is important to keep the operating theater clean and free from infection. Surgical gowns and masks are a possible carrier of bacterial infection to the patients. Hence, gowns and masks should be infection free and also act as a barrier to minimize the release of pollutant particles into the air. Conventionally, the gowns and masks are made from woven materials, largely cotton, that are both a source of dust, and, can easily release contaminated particles (Hutmacher, 2001). These textiles are lightweight, nonallergic, and have a high level of air permeability.

6.2.3 Extracorporeal devices

Fibers and textile technology is used in mechanical devices for purification, filtration, and circulation of blood from, for example, artificial livers or kidneys, or for artificial respiration. These devices provide support for vital organs through mechanical assistance. Artificial organs such as kidneys, livers, hearts, and lungs are made largely from cellulose, polyester, polypropylene fibers, or silicon membranes to mimic the natural function of the organ (Meena et al., 2013). Some intelligent textiles are used as mechanical devices that are designed to monitor body parameters and analyze the physiology of the human body. The majority of current smart textiles use physical sensors to monitor parameters such as movement (accelerometers), foot pressure, and breathing movement temperature (thermistors), and electric field and electric signals from heart and skeletal muscles (electrocardiography and electromyography respectively). Chemical sensor-based textiles are relatively new devices with emerging potential to monitor the health and well-being of the consumer through simultaneous interactions with body fluids or odors (Lay-Ekuakille, 2010).

6.3 Medical textile performance testing

Medical textiles are classified in Scheme 6.1. This scheme detailing constituents of various polymeric substances (eg, natural or synthetic) and their construction processes as outlined in the previous section is important in defining medical textiles aptly. Similarly, testing the properties and performance of medical textiles with suitable methods under accredited standards is essential. Thus, it is worthwhile to adequately describe all test methods, performance testing features, results analysis, and presentation guidelines of test results prior to marketing and commercialization of a device or product. This is done so that care and quality assurance of medical textiles will be documented appropriately, which benefits users.

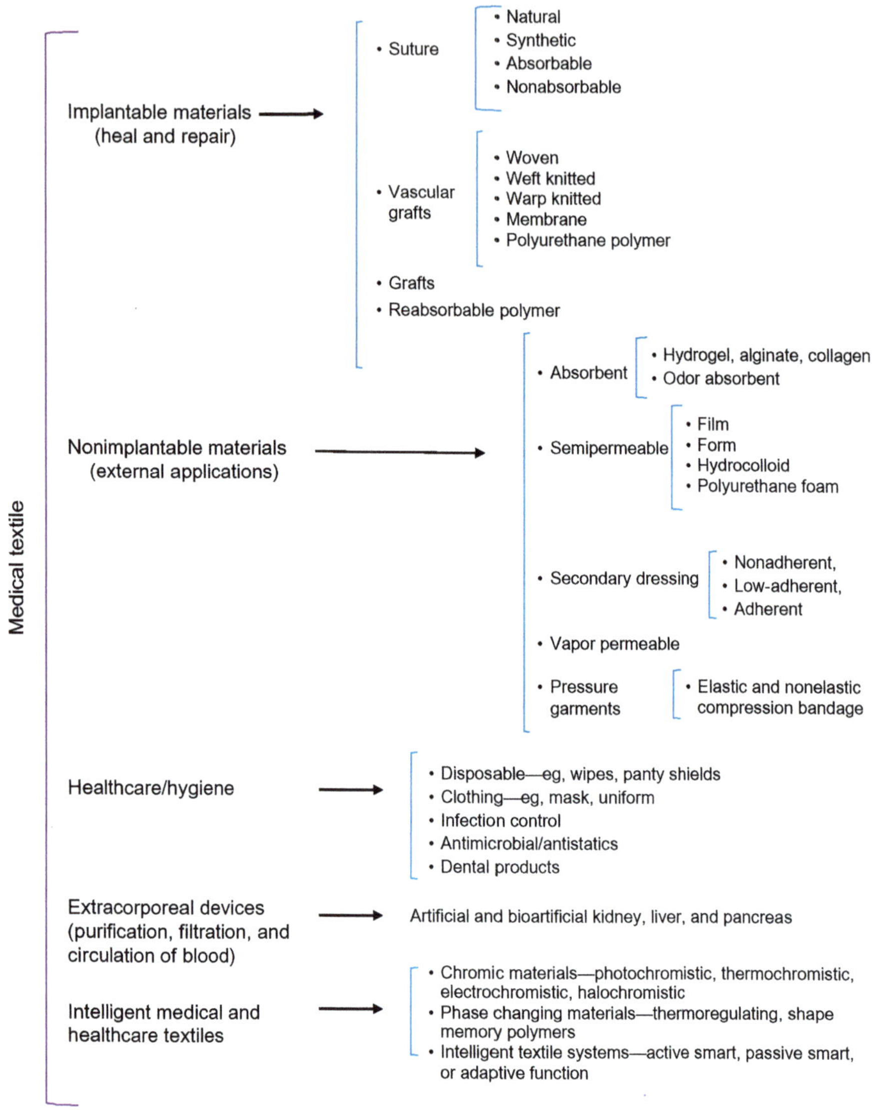

Scheme 6.1 Classification of various medical textiles (Rajendran and Anand, 2002b, 2012; Islam Kiron, 2015; Buschmann et al., 2015; Anand et al., 2005; Meena et al., 2013).

6.4 Methods, standards and validation

In response to ever-increasing innovation and development of medical textiles, medical textile performance testing is required during the manufacturing process. The test requirements developed by the European Medical Devices Directive are presented in this section. There are different classes of test methods and standards. However, the conformity with each test standard depends on the types and classes of the medical textiles. Moreover, the devices or products should satisfy and/or be capable of adhering to the rules and regulations of the European Union (EU). In 1993, the European Council Directive 93/42/EEC was published to establish requirements for all medical devices. It is often referred to as the Medical Device Directive (MDD), and is mandatory in the EU. This standard covers all medical device performance testing, including medical textiles, which describe the test requirements that need to be considered in a broad range of applications. However, in 1998, the European committee mandated the 'European committee for standardization (CEN)' to establish a European standard (EN 13795), more specifically for medical textiles, eg, surgical gowns, drapes, and clean air suits. This standard fills the missing technical part of the MDD. The EN 13795 demonstrates vital progress for better safety and provides better protection for people. The EN 13795 covers different aspects of the product from design, development, and production to their safe use condition, storage, and packaging, including transportation, labeling, and other physicochemical and microbial properties of the medical device. The product application considers the related risks, and each device is typically classified into Class I, Class II, or Class III. Class I covers low-risk devices, such as medical textiles that are used in operation theaters, which include gowns, drapes, and clean air suits. This standard has been written with a test method purpose, and aims to prevent the transmission of infection and contamination between medical staff and patients during surgical operations, and other potential hospital-related infections. The devices that possess a higher-risk potential for consumers, patients, and healthcare staff are classified under Class II or Class III. For example, medical textiles used for wound dressing are noninvasive devices, and typically fall in Class I, unless they are intended for wounds that breach the dermis. Wound dressings that manage the microenvironment of the wound, or are used for healing or repairing damaged skin tissues, fall in Class II (IIa and IIb). Similarly, medical textiles which are used either as nonimplantable devices (eg, antimicrobial agents consisting of dressings or patches produced from fibers or fabrics) or implantable devices (eg, cardiovascular grafts) are categorized in Class II. This is because when antimicrobial agents are applied on a fiber or fabric (textile), they are considered a drug, and the resulting device that includes the antimicrobial drug will be considered a combination product as defined in 21 CFR 3.29(e) in the EN 13795. In addition to monitoring and controlling health risks, this standard aims to provide a performance and safely level for the products during its entire lifecycle. The EN 13795 specifically focused on woven and nonwoven medical textiles and provides instruction for the test method required for the products. To date, the EN 13795 has been modified five times, the latest in 2013, and includes: BS EN 13795-2:2004+A1:2009, BS EN 13795-3:2006+A1:2009, BS EN 13795-1:2002+A1:2009. The latest version sets the minimum requirement for a product to be in compliance with the standard. The minimum value sets in the standard describe and distinguish the requirement for products for standard (less critical) and higher (critical) performance. The products within the critical zone are more involved with the chance of transferring the infection in the operation room, eg, the pieces of the surgical gowns which are in front and in the immediate zone of the operation site. On the other hand, products within a less critical zone are used, for example, to apply pressure during the surgical process or to be in contact with biological fluids.

6.4.1 Medical textiles performance testing (EN 13795)

The EN 13795 generally consists of three parts. The first part (EN 13795-1) consists of general requirements for production, and defines the design, processing, and assessment requirements of the products. This part aims to provide the same level of safety for the product through its entire lifecycle. In part 2 (EN 13795-2), the test method for assessing the characteristics of the product indicated in part 1 is described. The performance requirements of the products are described in part 3 (EN 13795-3) and are subdivided into critical and less critical zones for the products. High-risk products are more likely to be involved in infection transfer, and therefore, there should be clear labeling on the product to indicate the product's use in either the critical zone or less critical areas. The EN 13795 does not present specific suggestions for specific products that need to be used for surgical operation. However, risk assessment needs to be performed by the healthcare service provider, after asking the supplier of the product for information. For products intended to be sterilized or reused, the manufacturer should be able to provide enough information, eg, cleaning, packaging, and sterilization protocols, and product reuse techniques. Additionally, the product needs to have a clear label to show the intended application of the product zone area (high risk or low risk). The manufacturer should also provide test results and the rationale used for the differentiation and classification of the product as low or high risk. Moreover, before commercialization of the product, the manufacturer should test the finished product according to the guidelines provided by the standardization (validation) authorities such as the European Standards (EN), British Standards (BS), American Society for Testing and Materials (ASTM), International Organization for Standardization (ISO), and United States Food and Drug Administration (FDA).

Hence this chapter mainly explains the medical textiles performance testing methods recommended in part 2 of EN 13795 e.g.

6.4.2 Resistance to microbial penetration in dry (ISO 22612) or wet conditions (ISO 22610)

The resistance to microorganism penetration test quantifies the total microorganism counts that passed through a fabric when the fabric is in dry or wet condition. For example, contaminated talcum powders are poured and then plunged onto the fabric surface by using a metal plunger (Fig. 6.1). The results are reported as the number of colony-forming unit bacteria counted on an agar plate. The spores of *Bacillus subtilis* are normally used for this test. In the wet condition of this test (Fig. 6.2), the fabric is placed on a plate of agar, and the contaminated liquid (instead of talcum powder) is poured on the fabric. Typically, the test material is covered by a polyethylene film (10 μm thickness) then the contaminated fluid is subjected to a rotational mechanical abrasion stress using a mechanical finger under defined pressure (20 kPa, kilopascal) and time (15 min) so that the fabric is in contact with liquid and agar. The test is repeated five times in total for 15 min each. Then the CFU (colony-forming units) are observed after incubation at 35°C for 24 h, and results are reported as the BI (barrier index) in Table 6.1.

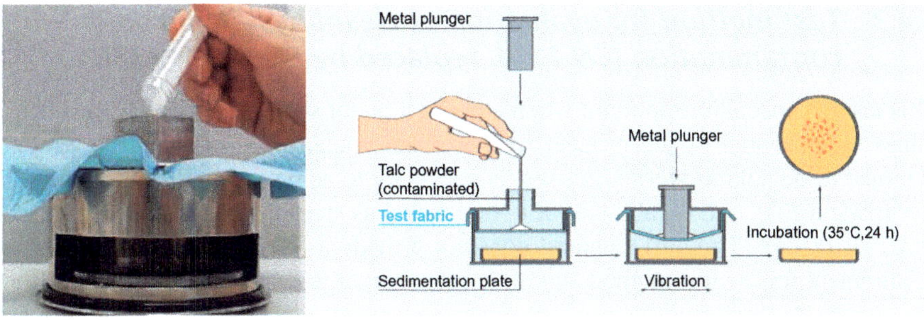

Fig. 6.1 Test device to evaluate resistance to microbial penetration in dry conditions.

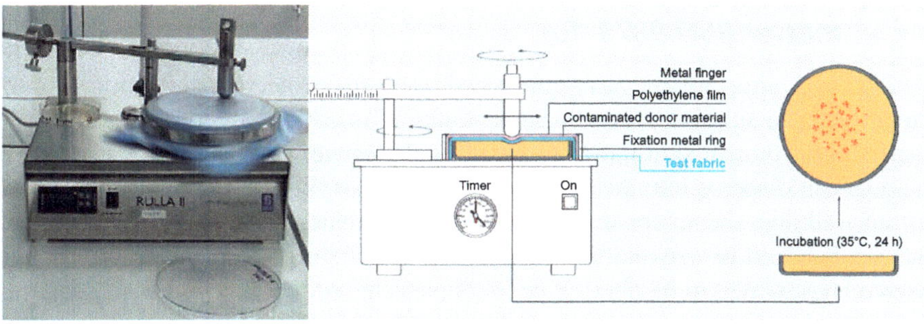

Fig. 6.2 Test device to evaluate resistance to microbial penetration in wet conditions.

Table 6.1 Summary of the performance requirement for textiles to microbial penetration

Unit	Standard performance		High performance	
	Critical area	Less critical area	Critical area	Less critical area
Surgical gowns and drapes				
Dry (Log_{10} (CFU))	N/A	≤300 CFU[a]	N/A	≤300 CFU[a]
Wet (BI)	≥2.8[b]	N/A	6.0[c]	N/A

[a]Test concentration: 10^8 CFU/g of talcum powder and vibration time of 30 min.
[b]The least significant differences for barrier index was found to be 0.98 at 95% confidence interval, therefore, materials varying by less than 0.98 BI are not different while materials varying by more than 0.98 are probably different.
[c]BI = 6.0 is the maximum achievable value and means no penetration for the purpose of this standard.

There are many ways textiles can be contaminated, due to their uses and/or through their environment. Potential threats include germs and fungi. However, there are a number of antimicrobial and antifungal medical textiles available nowadays. Antimicrobial performance testing using appropriate standard methods, test results, validation, and documentation is important. These tests are described in the following section.

6.4.3 Test method for evaluation of cleanliness—microbial contamination (EN 1174, replaced by EN ISO 11737)

This test is especially important when a product is not sold as sterile. The test determines the potential of the product for microbial contamination. When the product is aimed to be sold as sterile, this test is necessary to achieve the assurance level for sterilization (CFU:10^{-6}), allowing it to be labeled as sterile. The test results reported as CFU/100 cm^2. This test standard does not provide a fixed test procedure, but the requirement for the test method and the mechanism of test specified.

6.4.4 Test method for evaluation of cleanliness—particulate matter (ISO 9073-10)

Medical staff are generally very concerned about the protection and comfort of the patient and the surgical team. Because routes of infection are contact or airborne, dispersion of human skin particles are often the carrier of infection. A healthy individual can disperse into the air ~5000 bacteria-carrying skin particles per minute during walking, and males disperse more than females (Abreu, 2014). The particles are 5–60 μm in size and the average number of aerobic and anaerobic bacteria carried is estimated to be about 5 per skin particle (Abreu, 2014). For the determination of cleanliness—particular matter particle counts should be in the range of 3–25 μm, as particles in this size range carry microorganisms. The particle count result is reported as a particulate matter index (IPM) in Log(10) of particulate matter: IPM = Log(10)PM.

6.4.5 Test method for assessment of resistance to liquid penetration (EN 20811)

The ability of medical textiles to withstand the penetration of liquid contaminants is important for the health and safety of the healthcare staff and the patient (Behera and Arora, 2009; Kramar, 1998). Water spray impact tests evaluate the splash resistance of a fabric to liquid penetration. This test, also known as the hydrostatic test, evaluates the liquid penetration property of the fabric under hydrostatic pressure, and helps predict how the fabric will perform against fluid penetration. Considering that many procedures deal with the irrigation of tissue fluid, a fabric with high fluid resistance can be a great help by providing the required barrier against fluid penetration under applied pressure, for example, when a patient's arm comes in contact with contaminated fluid. For this test, the area of the test specimen should be 100 cm^2 with a water pressure increase rate of 10 ± 05 cm/min. Results expressed in a water column (cm) with higher values indicate better resistance of the tested fabric (Table 6.2). According to ASTM F1670 and ASTM F1671, body fluid or blood-borne pathogen stimulants should be used to perform the test while in AATCC 42 and 127; water is recommended for both spray and hydrostatic liquid penetration tests.

Table 6.2 Summary of the performance requirement for textiles to liquid penetration

Unit	Standard performance		High performance	
	Critical area	Less critical area	Critical area	Less critical area
Surgical gowns				
cm (H_2O)	≥20	≥10	≥100	≥10
Surgical drapes				
cm (H_2O)	≥30	≥10	≥100	≥10

6.4.6 Test method for measuring the linting of textile materials in the dry state (ISO 9073-10)

For this test, the fabric is subjected to cyclic torsion and axial efforts, similar to what is routinely practiced by healthcare staff (Fig. 6.3). The number of particles released from the fabric are measured in the same manner as the cleanliness test. However, a linting test measures the particle reserve in the product for a longer term. The result typically reports as the coefficient of linting in Log(10) values and calculates for the particle sizes of 3–25 μm.

Unit	Standard performance		High performance	
	Critical area	Less critical area	Critical area	Less critical area
Surgical gowns and drapes				
IPM	≤3.5	≤3.5	≤3.5	≤3.5
Log_{10} (lint count)	≤4.0	≤4.0	≤4.0	≤4.0

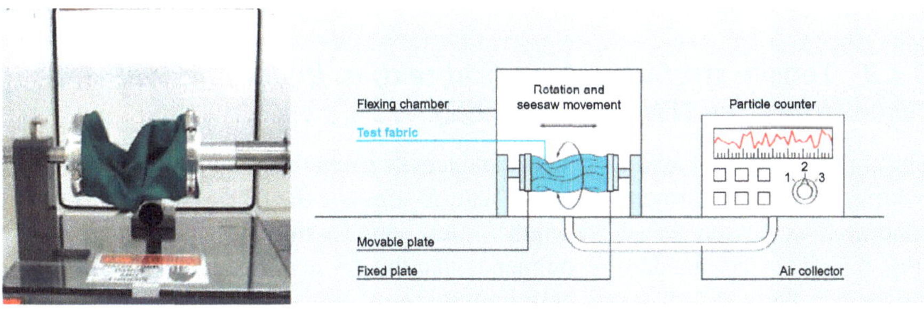

Fig. 6.3 Test device for measuring the linting of textile materials in the dry state.

6.4.7 Dry and wet burst strength (EN 13938-1)

The resistance of medical textiles to puncture or bursting is tested using the same method in dry and wet conditions (Fig. 6.4). This test is recommended because some area of the fabric is subject to bursting, or to puncture stress during use in the operating theater. The fabric can be damaged by the surgeon's elbow during an operation (DuPont Medical Fabrics, 2014). This damage can happen anytime, but it typically appears by exerting pressure on small areas of the fabric. A specifically designed apparatus is typically used to execute the test, which is capable of producing a various constant rate of increase in volume per unit time between 100 and 500 cm^3/min to within ±10% of the indicated value, or the apparatus with a testing time to burst of (20 ± 5) sec may be applied. The results of this test are expressed and reported in kPa and a higher reading indicates higher resistance of the fabric.

	Standard performance		High performance	
Unit	**Critical area**	**Less critical area**	**Critical area**	**Less critical area**
Surgical gowns and drapes				
Dry (kPa)	≥40	≥40	≥40	≥40
Wet (kPa)	≥40	N/A	≥40	N/A

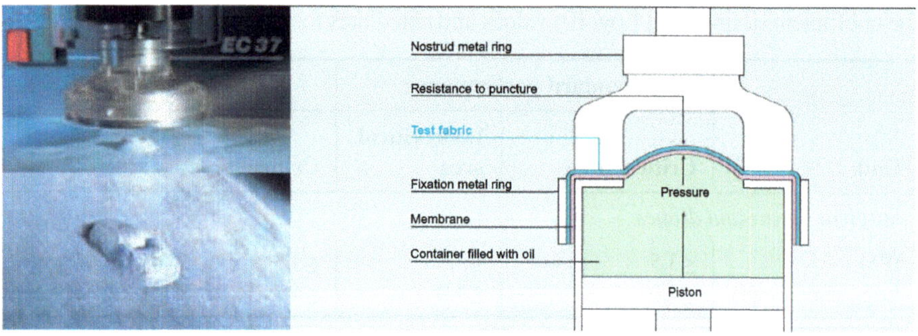

Fig. 6.4 Test device for examining the bursting strength of textile materials.

6.4.8 Tensile strength of medical textiles in dry and wet condition (EN 29073-3:1992)

The ability of medical textiles to withstand tensile strength is evaluated using this test. Normal wear of the garment imposes a natural stress on the fabric. This test is usually conducted to measure tensile strength for test samples in dry and wet environments (Fig. 6.5). This test standard recommends that the dry test specimens should be conditioned in the standard temperate atmosphere of a relative humidity of (65 ± 2)% and a temperature of (20 ± 2)°C for at least 24 h. The dimensions of test specimens are (50 ± 0.5) mm wide and of sufficient length to allow a jaw separation of 200 mm. Adequate length is required to avoid any risk due to heterogeneity of test specimen. At least five test specimens are required for dry testing, and another five test specimens are required for wet testing. To create the recommended condition for wet testing, the test specimens should be soaked with distilled water or fully deionised water for at least 1 h at (20 ± 2)°C. Then, the specimen is removed from the water, shaken off

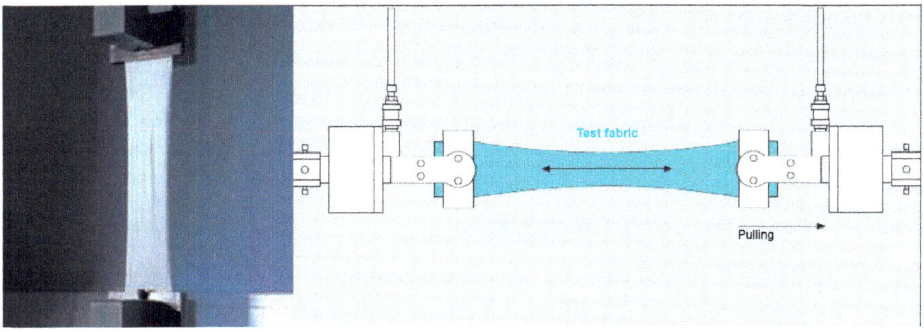

Fig. 6.5 Test device for examining the tensile strength of textile materials.

Table 6.3 Summary of the tensile strength performance requirement for textiles in dry and wet condition

Unit	Standard performance		High performance	
	Critical area	Less critical area	Critical area	Less critical area
Surgical gowns				
Dry (N)	≥20	≥20	≥20	≥20
Wet (N)	≥20	N/A	≥20	N/A
Surgical drapes				
Dry (N)	≥15	≥15	≥20	≥20
Wet (N)	≥15	N/A	≥20	N/A

excess water, and tested immediately. The results (Table 6.3) are reported in Newton (N), and higher numbers indicate better strength of the fabric.

6.4.9 The requirement and importance of medical textiles performance testing for the healthcare staff and the medical textile industry

The EN 13795 requirements implemented important modifications for better practices in the medical industry, which has led to highly sophisticated materials and processing systems and a higher degree of safety for patients and healthcare staff. Sophisticated guidance is provided for hospitals, fabric manufacturers, scientists, and purchasing organizations that are active in the medical textile industry. Generally, this standard provides better safety and protection for patients, nurses, surgeons, and other healthcare staff, and protects them against infections that could result from post-operation processes. Furthermore, this standard provides quality assurance and regulations for hospitals purchasing the products. The regulations in the standard set higher goals for manufacturers with specific and desirable criteria for improving devices and product quality, and also provide guidance for innovation. The textile industry, in particular the manufacturers involved in producing raw or finished materials that are used in manufacturing medical textiles (devices), is also required to set new quality control techniques to be consistent with the standard. For example, a reusable medical textile product manufacturer should specify the lifetime of the product, and the suitability of the product for reuse.

6.4.10 Tests for surgical masks

One in twenty hospitalized patients experience healthcare associated infections, which is approximately equal to 4 million infections and 37,000 deaths a year in the EU. Pathogen transfer normally occurs during the surgical procedure in operating rooms and other medical installments. The noses and mouths of the healthcare professionals on the operating teams are one of the main important sources of spreading and transferring the pathogens. Breathing, coughing, speaking, or sneezing can release different amounts of droplets from mucous membranes of nose and the mouth. These released droplets are dried, and the nuclei are suspended in the air, which can be transmitted to a susceptible site such as an open wound or sterile equipment. A surgical mask is the main barrier against this infection. The mask covers the nose, mouth, and chin, and provides a barrier for transmittance of the infections. Therefore, it is important to evaluate the performance efficiency of the mask for the protection against pathogen transfer. In the EU, the surgical mask should comply with the EN 14683 standard and have a CE label. This standard provides a requirement for design, construction, and testing methods for evaluation of the surgical masks.

Masks are classified into four types (see the following table) based on their performance efficiency, and, according to this standard, three tests are recommended as follows:

Types	Particulate respirator high fluid resistance (mmHg)	Particle filtration efficiency	Breathability—delta P (mm H_2O/cm^2)
Level 1	80	≥95%	<4.0
Level 2	120	98% at 0.1 μm	<5.0
Level 3	160	98% at 0.1 μm	<5.0
Level 4	160	99.9% at 0.1 μm	>5.0

6.4.10.1 Bacterial filtration efficiency

This test's objective and scope are to measure the amount of pathogens released through the mask in the air by measuring the amount of retained infection agents in the mask. According to the product's (mask) efficiency, the product is classified in Type 1 (95%) and Type 2 (98%). Over several years, the bacterial filtration efficiency (BFE) has had different requirements as established by the European committee, which were used to amend EN 14683 and ASTM F2101 (in the case of the United States). Therefore, manufacturers have to comply with both requirements by performing two separate tests. This was due to differences in the preconditioning requirement by each test as outlined in Table 6.4. However, in the latest revision of the EN 14683, the preconditioning

Table 6.4 **Preconditioning requirements**

	After EN 14683
Standard	Preconditioning requirement: EN 14683:2014
EN 14683	21 ± 5°C and 85 ± 5% relative humidity prior to testing
ASTM F2100	21 ± 5°C and 85 ± 5% relative humidity prior to testing

requirement of both tests aligned and the manufacturer needed to only use one test. ASTM 2101 does not define the acceptable level for bacterial deficiency, and therefore, is required to report the specific condition under which the test is conducted. For this test the *Staphylococcus aureus* is used as the challenging microorganism to evaluate the filtration efficiency. The mask is challenged by subjecting it to bacterial aerosol at a flow rate of 28.3 L/mm (1 ft^3/min) inside the medical face mask (Fig. 6.6). The test can be performed using both sides of the mask (face side and liner side) to evaluate the filtration efficiency of the mask related to aerosol generated from both the patient and the wearer of the mask. Testing the mask without considering the parameters such as physiochemical degradation, thermal stress, or wetting contaminant might lead to an inaccurate result, and consequently a false sense of security; therefore, masks need to be tested after suitable pretreatment if the stress conditions are of concern. The test method is quantitative and the maximum filtration efficiency that can be determined using this method is 99.9%. The ASTM 1201 does not measure the breathability of the mask or any other items related to ease of breathing using the medical face mask.

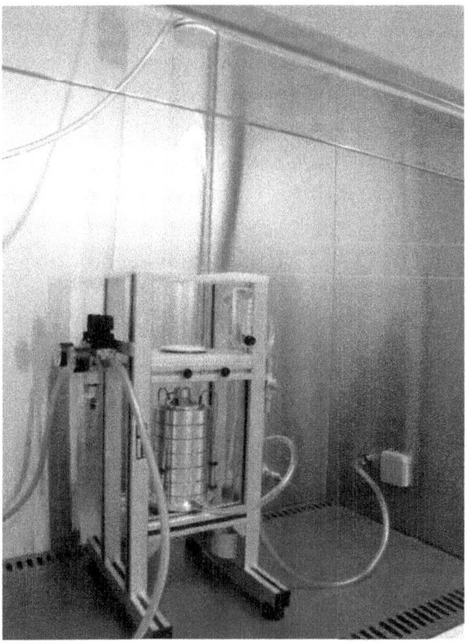

Fig. 6.6 Test instrument and set-up of measuring of BFE.

Also, it needs to be considered that this test only evaluates the filtration efficiency of the material used for construction of the mask and does not evaluate design, fit or facial sealing abilities of the medical face mask. Similarly, other medical textiles (garments), eg, surgical gowns, surgical drapes, and sterile barrier systems, can also be tested for BFE using this test.

Typically the resistance of mask to airflow is determined in this test as:

Type 1 and 2 (nonsplash resistant) $\leq$29.4 Pa/cm^2
Type IR and IIR (splash resistant) $\leq$49.0 Pa/cm^2

6.4.10.2 Splash resistance (synthetic blood)

Blood vessels can be punctured or damaged during medical procedures, which results in a blood splash that requires a face mask to resist this impact. Several parameters are involved in the splash of blood from the punctured vessel, including the size of the puncture and the distance from the puncture. Medical face masks should be able to resist liquid penetration (eg, blood splash, body fluids). Parameters such as viscosity, polarity, and surface tension of the fluid are important factors affecting the splash resistance of the mask. On the other hand, the structure, hydrophilicity, and hydrophobicity of the material and the design of the mask also are crucial parameters affecting the resistance of the mask. For this test, the synthetic blood is prepared with a red dye and surface tension of the synthetic blood is adjusted 0.042 ± 0.002 N/m. However, this blood will not always duplicate the polarity and wetting behavior of the real blood. This test assumes that the medical face mask will be in a close distance of ~300 mm to the splash area, and a blood stream with velocities of 80, 120, and 160 mmHg are used for this test (Fig. 6.7). These blood velocities cover the mean human blood pressure of 80–120 mmHg. The maximum value is 120 mmHg, which corresponds to average systolic blood pressure. The mask is intended to protect against small splashes of blood from the rupture of small arterials. In addition to the three preceding tests, the FDA (Food and Drug Administration) also prescribes the measurement of filtration efficiency regarding inert particles (latex), fire tests, and tests to measure biocompatibility. Therefore, the manufacturer should be able to test and classify their surgical masks in type, for the European market (EN 14683) and for the American one (ASTM F2101). The precondition requirements for the test are described in Table 6.4.

While the preconditioning requirements aligned, the required tests for both standards and the performance results are different between two standards, which typically depend on the claimed properties of the mask (Table 6.5).

In addition to standard EN 13795, there are a number of other standards for medical textiles and surgical masks proposed and described in the EN standard, ASTM, and also in international standards, including standards for testing of water vapor permeability (breathability) of the fabric in EN 31092 and ISO 11092, thermal insulation testing of the fabric in EN 15831 and ASTM F2370, air permeability of the fabric in ISO 9237, water penetration resistance in ISO 811, nontoxicity of the medical device in ISO 10993, measurement of the electrical risk in ISO 2878 (BS2050) and EN 1149, protection against laser beams, primary ignition, and penetration in ISO 11810-1 and

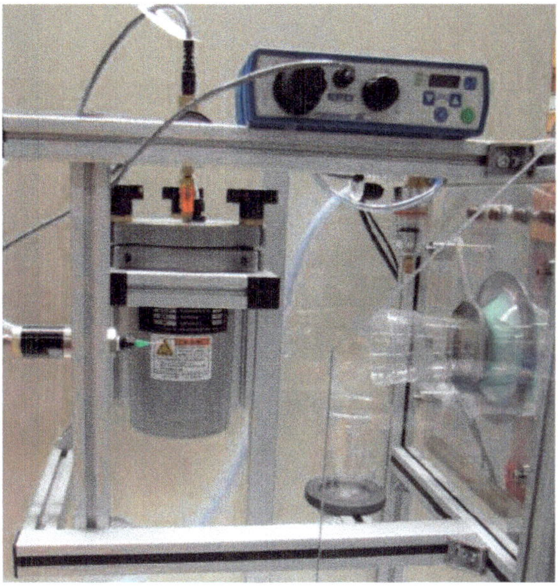

Fig. 6.7 Test instrument and set-up of measuring splash resistance.

secondary ignition in ISO 11810-2. These test standards are important and required to validate the performance efficiency of the medical textiles.

Water penetration resistance (ISO 811)

Breathability is the ability of the fabric to permit water vapor to pass through and to prevent the entry of water (Behera and Arora, 2009). The comfort properties of a fabric depend on its ability to transmit water and vapor from the body to prevent accumulation of liquid on the skin. In this way, thermal energy generated by the body will be transmitted, and vapor moisture will be diffused, resulting in a comfortable condition (McCann and Bryson, 2014). The value of breathability is calculated by measuring the quantity of moisture vapor through a fabric during a specified time period and reported as the moisture vapor transmission rate (MVTR) (Gokarneshan et al., 2015). Higher values indicate better removal of vapor and moisture, and prevention of the accumulation of gas and vapor. The MVTR is reported as grams per square meter (g/m^2) under defined temperature and humidity (Gokarneshan et al., 2015). Air permeability is another factor that affects the comfort property of the fabric and has a close relation to water vapor permeation. The ability of the fabric to permeate air or water might change due to changes in the different temperature and humidity conditions, eg, the fabrics made of cotton and wool might experience resistance to air or water permeation due to swelling and closeness of the pores in the fabric. A surgical gown needs to be completely waterproof, and considering a moderate working speed of 20°C, the gown needs to provide the air permeability of 100 L/m^2/min and the water vapor permeability of 400 g/m^2/24 h. The resistance of the fabric to pass the water is measured by subjecting materials (100 cm^2) to a steadily increasing pressure of water

Table 6.5 **Test requirements and performance requirements by Type, in EN 14683:2014 and level barrier in ASTM F2100-11**

	Characteristic					
Test	Type I	Type II	Type IIR	Level 1	Level 2	Level 2
21 ± 5°C and 85 ± 5% relative humidity prior to testing	≥95	≥98	≥98	≥95	≥98	≥98
Differential pressure, mm H_2O/cm^2						
H_2O/cm^2	<3.0	<3.0	<5.0	<4.0	<5.0	<5.0
Pa/cm^2	<29.4	<29.4	<49.0	<39.2	<49.0	<49.0
Submicron particulate filtration efficiency at 0.1 micron (%)	Not required	Not required	Not required	≥95	≥98	≥98
Splash resistance/synthetic blood resistance (mmHg)	Not required	Not required	120 (16.0 kPa)	80	120	160
Pass result						
Flame spread	Not required	Not required	Not required	Class 1	Class 1	Class 1
Microbial cleanliness (CFU/g)	≤30	≤30	≤30	Not required	Not required	Not required

on one face until water passes. The pressure at which penetration occurred is noted. The material should be clamped during the test period without slipping, and no tendency should occur at the clamped zone. For this test, only distilled water at 20 ± 2°C should be used, and the pressure should increase 10 ± 0.5 cmHg or 60 ± 3 cm H_2O/min and the results obtained by both rates should be reported, and the pressure should be read at 0.5 accuracy cm H_2O.

Thermal manikin (EN ISO 15831—ASTM F2370)

This test is used to measure the thermal insulation of the fabric. It is possible to measure the breathability and define the range of the fabric utility when the test is combined with Skin Model (Huang and Qian, 2008). The evaporation resistance of different clothing systems can be quantified using this test under isothermal conditions. The test is performed using a manikin, which is dressed completely; then the manikin is placed into a chamber with an adjustable climate condition, different temperature, and variable humidity and wind speed. The obtained data then can be used to model and predict the physiological response of the wearer in different climate conditions (OECD, 2009a). Considering that there are many parameters involved in this test, the measurement of evaporation resistance is a complex phenomenon. The measurement includes heated body parts such as feet (EN 345), hands (EN 511), gloves to protect against the cold, and whole body manikins (ASTM F1291). The standard for measuring the thermal insulation of clothing uses a heated thermal manikin (ASTM F2370-05) and determines water vapor resistance using a sweated thermal manikin (ASTM F1291-05). Furthermore, ISO 15831 provides methods for determining clothing insulation using a heated thermal manikin.

Cytotoxicity and nontoxicity performance testing (ISO 10993)

When the medical textile or part of it is considered as the medical device, then its toxicity performance testing requires validating under the ISO 10993 test standard. The fabric or parts of it are put in either direct or indirect contact with the cell culture system (eg, L929 cell line), and then the cell viability results determine the release of toxic materials from the tested fabric or device.

The in vitro tests for cytotoxicity assess the response of cells in culture to direct contact with devices or to their extracts. The ISO 10993-5 (1999): "Tests for In Vitro Cytotoxicity" specifies procedures for testing devices by direct or indirect contact, extracts of devices, and filter diffusion. The test should be performed on: (a) an extract of the test sample, (b) the test sample itself. Extracts of test devices and materials (eg, medical textiles) are tested by exposure to the cell culture system (eg, L929 mouse fibroblast cell line). The presence of cytotoxic leachates is indicated by loss of cell viability. In the direct method, the test device or a part of it is directly exposed on a mammalian cell layer which is protected by an agar layer. In this way, the cytotoxic leachate diffuses through the cell layer through agar, and the loss of viable cells indicates the toxicity of the tested material. In the direct contact method, the test material is directly put on the cell culture system, and no protective agar layer is used. About a half million to one million of cells are used for each test, and cytotoxicity of the material is measured after 24–72 h of cell exposure to the extract or the material. A useful way to grade test samples is given in Tables 6.6 and 6.7.

Table 6.6 Grading recommended for in vitro cytotoxicity test results

Grade	Reactivity	Conditions of all cultures
0	None	Discrete intracytoplasmic granules, no cell lysis, no reduction of cell growth
1	Slight	Not more than 20% of the cells are round, loosely attached and without intracytoplasmic granules, or show changes in morphology; occasional lysed cells are present; only slight growth inhibition observable
2	Mild	Not more than 50% of the cells are round, devoid of intracytoplasmic granules, no extensive cell lysis; not more than 50% growth inhibition observable
3	Moderate	Not more than 70% of the cell layers contain rounded cells or are lysed; cell layers not completely destroyed, but more than 50% growth inhibition observable
4	Severe	Nearly complete or complete destruction of the cell layers

Table 6.7 Cytotoxic test procedure recommended by ISO 10993-5

ISO 10993-5
0.5–1 million of cells per dish
Extract ratio:
If the thickness ≥1.0 mm
25 cm^2 per 20 mL
If thickness ≥0.5 mm
60 cm^2 per 20 mL
If thickness ≤0.5 mm
120 cm^2 per 20 mL
24–72 h of exposure period
Toxicity determination: quantification of the survived colonies
Positive controls: materials that provide a reproducible cytotoxic response

Furthermore, some materials, including medical textiles, have the potential to cause allergy or irritation. The sensitization potential of the material is therefore required to be tested. The ISO 10993-10 assesses the possible contact hazards from chemicals released from the devices, which may produce skin, mucosal, or eye irritation and skin sensitization. Currently, there has been no satisfactory in vitro test to eliminate the requirement for in vivo testing to evaluate irritation. However, the rat skin Transcutaneous Electrical Resistance (TER) test and the human skin model test have been internationally validated and accepted as alternative tests to assess the skin corrosion with chemicals (OECD, 2009a,b). To perform the test, an animal (rabbit) with the healthy intact skin needs to be used and the fur on the back of animal is clipped 24–4 h of testing with approximate size of 10 cm × 15 cm and then 0.5 g or 0.5 mL of materials apply to the site and then the application site is covered with

Table 6.8 Scoring system for skin reaction

Reaction	Irritation score
Erythema and eschar formation	
No erythema	0
Very slight erythema (barely perceptible)	1
Well-defined erythema	2
Moderate erythema	3
Severe erythema (beet-redness) to eschar formation preventing grading of erythema	4
Oedema formation	
No oedema	0
Very slight oedema (barely perceptible)	1
Well-defined oedema (edges of area well-defined by definite raising)	2
Moderate oedema (raised ~1 mm)	3
Severe oedema (raised more than 1 mm and extending beyond exposure area)	4
Maximal possible score for irritation	8
Other adverse changes at the injection sites shall be recorded and reported	

Table 6.9 Primary or cumulative irritation index categories in a rabbit

Mean score	Response category
0–0.4	negligible
0.5–1.9	slight
2–4.9	moderate
5–8	severe

2.5 cm × 2.5 cm nonocclusive dressings (such as a gauze patch). The chemical characterization of the test materials is the prerequisite for this test. The irritation scores of all animals are the main measuring technique that are added together and divided by the total number of animals and reported as the cumulative irritation index (eg, Tables 6.8 and 6.9).

Human skin irritation test

According to ISO 10993-10, two of the most commonly used methods of testing for skin sensitization or irritation are the Guinea pig maximization test (GPMT) and the Buehler closed-patch test. Of these, the maximization test is the most sensitive method. The Buehler closed-patch test and the Magnusson–Kligman GPMT allow the expression of such allergic materials by an initial stimulation. Both tests can be used for the performance testing of medical textiles. However, the maximization test is recommended when the test material is manageable for intradermal injection. The

Table 6.10 **Magnusson and Kligman scale**

Patch test reaction	Grading scale
No visible change	
Discrete or patchy erythema	1
Moderate and confluent erythema	2
Intense erythema and/or swelling	3

sensitization is highly dependent on the applied dose, and if the irritation threshold is not reached, then the highest possible concentration might apply. However, the health of the animal should be considered, therefore, the induction dose is normally selected through preliminary tests. For the GPMT, a healthy young animal should be used, and in the case of powder or liquid materials, a minimum of ten animals should be treated with the test sample, and five animals should be used as the control group. For testing the extract of the material, again ten animals should be treated with the sample, and five should be used as the control group. The sample material is applied in the form of a patch 4–8 cm^2 and after 24 h the application site is assessed for erythema and odema using the Magnusson and Kligman grading system, which is shown in Table 6.10.

6.5 Care and quality assurances

Medical textiles in the healthcare or hospital care sectors play many important roles nowadays (Anand, 2001; Rajendran and Anand, 2012; Meena et al., 2013). As research has progressed, fibers and textile materials have found their way into a variety of healthcare and medical applications; as a result, medical textiles sectors are growing enormously. In medical and healthcare (hospital care) sectors, medical textiles and related products are used in the operating theater of the hospital ward for the hygiene, care, and safety of staff and patients (Rajendran and Anand, 2012; Meena et al., 2013). In many cases, these textiles are not yet defined aptly or well-known as medical textiles, but are used as hospital garments. However, at present, manufacturers, producers, and users are not getting enough information about the suitability of the textile materials (garments) used in hospital care and/or in surgery applications. Medical textiles as disposable products have been proposed for operating room garments and drapes with the objective of reducing microbial contamination of the incision and of protecting the operating room staff from infection. But the real practices are different, where reusable materials have been often used in surgery. This situation is reversed nowadays due to the spread of Ebola, the HIV virus, Hepatitis B infection, and other viruses. So the need to protect healthcare workers from the patient has become a major concern, which creates research opportunities and market demands for the development of more effective protective medical textiles. Furthermore, manufacturers or producers, retailers, and the services industry also need to consider the environmentally conscious population when deciding between recycling potentially contaminated products and/or disposing of the products in landfills.

By reviewing literature, market research reports, and accounting for recent advancements in medical procedures and textile engineering, the use of textile materials as medical textiles in the healthcare or hospital care industry is growing immensely (Marketsandmarkets, 2015). Hence, care and quality assurance of medical textiles are critically important to consider. Adequate actions have not yet been taken in ensuring quality and performance of medical textiles. At the same time, enough actions have not yet been taken to provide information associated with product performance testing standards to manufacturers or producers for ensuring product care and quality assurances. Thus, it appears that the product performance test results, validation, and correct documentation are important, and useful for everyone from manufacturers to end-users. With this information, manufacturers and retailers can demonstrate due diligence to authorities within the target markets through the certification of suppliers. Furthermore, the care and quality assurance documentation will allow the entire industry to gain consumers' trust through product safety reliance.

6.6 Medical textiles and its future trends

Medical textiles are one of the most important, and fastest-growing sectors of the textile industry. The medical textiles industry has been enhanced with existing products, and also creating new ones with new materials and innovative designs. Some of these new products are being designed for accelerating healing, infection control, and noninvasive surgical procedures. At present, the global market for medical textiles is worth hundreds of billions of dollars (Eriksson and Sandsjo, 2015; Robert and Rachel, 2015). The world demand for disposable medical supplies is forecasted to expand 6.2% annually to $198 billion in 2016 (Robert and Rachel, 2015). Increased enforcement of infection prevention standards, together with a growing number of hospital, surgical, and outpatient procedures, will enhance overall advances (Marketsandmarkets, 2015; Textile World, 2015; Shishoo, 2011; Eriksson and Sandsjo, 2015; Robert and Rachel, 2015). Population growth, aging populations, and the construction of new medical facilities are a driving force for this industry. Hence its importance will increase even more in the future. Medical textile products cover a wide range of uses, which include surgical gowns, operating room drapes, face masks, staff uniforms, hospital bedding and curtains, hygienic wipes and bandages, cardiovascular grafts, and also a variety of other unique applications. Medical textiles are increasingly used for hygienic purposes, and for first aid and clinical purposes, in addition to wound care practice such as protection from infection, bandaging, and applying pressure on wounds (Textile World, 2015; Shishoo, 2011). However, it appears that high energy costs, lack of raw materials, transportation costs, and unawareness in developing countries of hygienic practices might slow developments in the medical textile industry, and/or hinder the advancement of technology. The invention and development of new fibers and fabrics allow combining several functionalities such as breathability, temperature control, shock proofing, resistance to fluids, and a lot more in one material. However, the growth of care for the aging population has been broadening the market for new products. Treatment and management of diabetic wounds is a good example. Nevertheless, the incorporation of healing substances shifts medical textiles to innovative medical products (Rajendran and Anand, 2002a). Also, future research should be focused on development of suitable technology and standards to better meet the growing demand for health and safety products. Thus, the existing performance testing standards need to be refined and modified further through more experiments and data collection processes. In this way, some parameters such as breathability and water vapor transmission need to be determined objectively to obtain and specify the minimum comfortable limits. To achieve this, research and development is required to design and develop medical textiles with different levels of permeation, porosity, gas permeability, etc., through application of new raw materials and optimizing current techniques of material construction and coating (Gokarneshan et al., 2015). The biopolymer and genetically modified microorganism will be developed for production on new fibers as well. An example is polyhydroxybutyrate (PHB), developed by Zeneca Bio products. High molecular weight polyester with a bacterial source, which is biocompatible and biodegradable with suitable thermoplastic properties, can melt into fibers that are knitted into manufactured medical textiles (Plackett, 2011). Production and development of new fibers expands the range of applications of medical textiles even more, and therefore, the utilization of textiles for medical use becomes more challenging. However, the need to improve and understand the performance of these new materials is inevitable to target specific applications. Environmentally friendly processes such as enzymatic processes need to be applied more often and replace the chemical treatments of textiles, so that enzymes with low energy requirements and higher yields can be developed (Plackett, 2011). The quality control system and standards, therefore, need refinement to meet new waste management criteria and programs deemed to be essential. Nevertheless, the future of the medical textile market is more dependent on the development of innovative products with improved comfort, higher performance, and lower cost. Table 6.11 summarizes a developing stage of technical textiles, including medical and smart textiles.

6.7 Conclusion

Several pieces of literature and reports documented that the use of medical textiles in healthcare and hospital care sectors is growing significantly. This is largely due to the care for the quickly rising aging population, and the need to prevent the spread of various infectious diseases. Hence, the performance testing and care and quality assurance of medical textiles are critically important at the present time and in the future as well. In this chapter, some performance testing methods and standards for the evaluation of medical textiles are reviewed and discussed. In many cases, there is a consensus between the defined requirement by different organizations, including EN, ISO, and ASTM. However, medical textiles manufacturers are still required to meet performance testing requirements to be able to present their products in markets that

Table 6.11 **Development stages of technical textiles**

Time horizon (Years)	Barrier	Early warning	Sustainability	Functionality
0–3	Barrier maintained with lower basis weights at lower cost • Fully impervious to bacteria, virus, prions and fluids • Simpler material structures	None at present	Incorporation of some biopolymers at low additive levels • Down gaging and light weighting • Reduction of medical waste costs	• Water vapor transmission rate and breathability of fabric influenced by body temperature • Connectivity—gowns and gloves • Cooling or heating apparel and surgical drapes
3–5	Odor elimination • Smoke elimination • Tear-proof and puncture-proof drapes, gowns and gloves	Detection and disclosure of a break in barrier	Biobased polymers becoming economically justifiable • Move to green materials via current or green manufacturing technologies, reducing energy usage	• Monitoring capabilities—temperature, blood pressure, breathing, oxygen—through embedded sensors • Conformable materials for comfort, ease of donning and doffing apparel • Integrated radio-frequency identification tracking for compliance
5+	Active enhancement of the environment around the drape to promote bacterial and viral barrier	• Sensors fine-tuned to detect patient distress, providing feedback and alerts • Self-healing/repairing of breached materials	Self-cleaning surfaces in hospitals—operating room tables, curtains and cubicles • Reusable and disposable products • Resterilizable and reusable products	• Encapsulated solidifying technology that provides instant rendering of bodily fluids noncontaminating, both in mobility and potency

requires either of the test standards endorsed. Also, these performance testing methods (standards) that are required to ensure care and quality assurance, including reuse and resterilization, should be checked regularly, accounted for, and described.

6.8 Sources of further information and advice

Test methods or standards, images, and websites:

EN	European standards (European Committee for Standardization, CEN, Comite' Europee'n de Normalisation); http://www.cencenelec.eu/standards
BS	British Standards Institution; https://www.standardsuk.com
ASTM	American Society for Testing and Materials; http://www.astm.org/
ISO	International Organization for Standardization; http://www.iso.org/
FDA	United States Food and Drug Administration; http://www.fda.gov/

Textile world; http://www.textileworld.com
Centexbel, Textile Competence Centre; http://www.centexbel.be
NIOSH's Skin Exposures and Effects at http://www.cdc.gov/niosh/topics/skin
Quick Selection Guide to Chemical Protective Clothing, Fourth Edition, at http://www.cdc.gov/niosh/ncpc1.html
Council Directive 93/42/EEC of Jun. 14, 1993 concerning medical devices; http://eur-lex.europa.eu/legal-content/EN/TXT/?qid=1459226038342&uri=CELEX:31993L0042
OSHA Dermal Exposure Topic Page at https://www.osha.gov/SLTC/dermalexposure/index.html
http://www.fibre2fashion.com/industry-article/technology-industry-article/biotech-textiles/biotech-textiles4.asp
http://www.textileworld.com/Issues/2012/November-December/Nonwovens-Technical_Textiles/Medical_Textiles-How_Smart_Do_They_Have_To_Be
http://www.marketsandmarkets.com

Acknowledgments

Various standard test methods and associated test device (set-up) images are stated in this chapter. The authors are gratefully acknowledge the European standard authority, British standard authority, and American standard and testing authority.

References

Abreu, M.J., 2014. Suitable control choices: exploring the roles of clean air suits and scrub suits in infection control. In: EMH Medical Workwear Article. University of Minho, pp. 21–27.
Anand, S.C., 2001. Medical Textiles. Woodhead Publishing, Abington.
Anand, S.C., Kennedy, J.F., Miraftab, M., Rajendran, S., 2005. Medical Textiles and Biomaterials for Healthcare. Elsevier Science.
Behera, B.K., Arora, H., 2009. Surgical gown: a critical review. J. Ind. Text. 38, 205–231.

Buschmann, H.J., Dehabadi, V.A., Wiegand, C., 2015. Medical, cosmetic and odour resistant finishes for textiles. In: Paul, R. (Ed.), Functional Finishes for Textiles. Woodhead Publishing, pp. 303–330 (Chapter 10).

Czajka, R., 2005. The world textile industry is moving rapidly toward the manufacture of high-added value fibres, fabrics and textiles products. Fibres Text. East. Eur. 13 (1 (49)).

DuPont Medical Fabrics, 2014. EN 13795: Understanding the European Standard—Surgical Gowns, Drapes and Clean Air Suits.

Eriksson, S., Sandsjo, L., 2015. Three-dimensional fabrics as medical textiles. In: Advances in 3D Textiles. A Volume in Woodhead Publishing Series in Textiles, Woodhead Publishing, pp. 305–340.

Gokarneshan, N., Rachel, D.A., Rajendran, V., Lavanya, B., Ghoshal, A., 2015. Emerging Research Trends in Medical Textiles. Springer, Singapore.

Huang, J., Qian, X., 2008. Comparison of test methods for measuring water vapor permeability of fabrics. Text. Res. J. 78, 342–352.

Hutmacher, D.W., 2001. Scaffold design and fabrication technologies for engineering tissues—state of the art and future perspectives. J. Biomater. Sci. Polym. Ed. 12, 107–124.

Islam Kiron, M., 2015. Introduction of medical textile. http://textilelearner.blogspot.com/2012/02/introduction-of-medical-textiles.html#ixzz3rKfUm8le (accessed 08.09.2015).

Kramar, L., 1998. Recent and future trends for high performance fabrics providing breathability and waterproofness. J. Ind. Text. 28, 106–115.

Lay-Ekuakille, A., 2010. Wearable and Autonomous Biomedical Devices and Systems for Smart Environment: Issues and Characterization. Springer, Berlin.

Marketsandmarkets, 2015. Medical textile testing market by test (dimensional stability, toxicology, biocompatibility, microbiological & flammability), by product (gown, drape, wrap, uniform, curtain, bedding, & face mask), and by region—Global Trends and Forecasts to 2020. Marketsandmarkets research report (report code: UC1414). Published date: Nov. 2015.

McCann, J., Bryson, D., 2014. Textile-Led Design for the Active Ageing Population. Elsevier Science.

Meena, C.R., Ajmera, N., Sabat, P.K., 2013. Medical textiles. In: Textiles for Industrial Applications. CRC Press, pp. 39–100.

OECD, 2009a. Test No. 430: In Vitro Skin Corrosion—Transcutaneous Electrical Resistance Test (TER). OECD Guidelines for the Testing of Chemicals, Section 4: Health Effects. OECD Publications.

OECD, 2009b. Test No. 431: In Vitro Skin Corrosion—Human Skin Model Test. OECD Guidelines for the Testing of Chemicals, Section 4: Health Effects. OECD Publications.

Plackett, D., 2011. Biopolymers: New Materials for Sustainable Films and Coatings. Wiley.

Rajendran, S., Anand, S.C., 2002a. Development in medical textiles. Text. Prog. 10–13.

Rajendran, S., Anand, S.C., 2002b. Developments in medical textiles. Text. Prog. 32, 1–42.

Rajendran, S., Anand, S.C., 2012. Woven textiles for medical applications. In: Woven Textiles: Principles, Technologies and Applications. A Volume in Woodhead Publishing Series in Textiles, Woodhead Publishing, pp. 414–441 (Chapter 14).

Robert, C., Rachel, A., 2015. ITA Technical Textiles Top Markets Report.

Shishoo, D.R., 2011. High-performance textiles and nonwovens are targeted for growth. Specialty Fabr. Rev. https://secure.ifai.com/sfreview/articles/0311_wv_asia_potential.html.

Textile World, 2015. Textiles 2015: More Improvement Ahead. Jan./Feb. http://www.textileworld.com/textile-world/features/2015/02/textiles-2015-more-improvement-ahead/.

Walker, I.V., 1999. Nonwovens – the choice for the Medical Industry into the next millennium. In: Proceedings of Medical Textile Conference. Bolton Institute, UK Publishing, Cambridge, pp. 12–19.

Multiscale characterization and testing of function-integrative fiber-reinforced composites

7

S. Geller, K. Holeczek, A. Winkler, T. Tyczynski, T. Weber, M. Gude, N. Modler
Dresden University of Technology, Dresden, Germany

7.1 Introduction

Extensive efforts on lightweight construction lead to increasing application of fiber-reinforced polymers (FRP) in various industrial fields. In this context, the implementation of smart composites is gaining importance and is typically connected with the integration of sensors and actuators. Numerous research and development activities focus on functionalization of composite structures by piezoelectric transducers (eg, Macro Fiber Composites (MFC), Active Fiber Composites (AFC) or DuraAct) for various tasks like condition monitoring, damage detection, or even for active vibration damping (Melnykowycz et al., 2005; Lin and Chang, 2002; Edery-Azulay and Abramovich, 2006). Initial prototypic applications have been realized by adhesive bonding or laminating these functional elements into thermoset based composites (Konka et al., 2012). Nevertheless, FRP processing technologies offer outstanding design flexibility combined with moderate processing conditions, which allow for material homogenous embedding of adapted piezoelectric modules (Hufenbach et al., 2009). By this, smart fiber-reinforced composites can be manufactured by combining part manufacturing and module integration in single-stage efficient processes capable for series production. In the following, various qualitative and quantitative methods for the multiscale characterization (from filament to component) of such smart fiber-reinforced composites with respect to their electromechanical behavior are shortly described. By example of case studies, specific application scenarios are presented.

7.2 Common techniques for the characterization of smart fiber-reinforced composites

Since a system is only as strong as its weakest link, every subsystem of a smart composite needs to be inspected in order to assure reliable operation of the whole assembly during its entire in-service lifetime. Therefore, the physical properties of all constituents of the smart composite, the mechanical and environmental behavior, as well as the intended functionality, have to be characterized and checked against technically reasonable references on every assembly step.

Performance Testing of Textiles. http://dx.doi.org/10.1016/B978-0-08-100570-5.00008-6

In general, smart structures can be considered as a compound of following subassemblies:

- Fiber and matrix material of the composite component assuring its load-bearing capabilities,
- Sensor elements essential to measure signals connected with the aimed functionality,
- Actuator elements necessary if the smart structure is designed to influence either its own behavior or the surrounding,
- Electronic components mainly required to transfer signals from the sensors and to the actuators as well as for power supply and signal conditioning,
- Software modules to interpret the measured signals in such a way that the aimed functionality is reached.

Such complex design of smart composite structures connected with the testing necessity of every subsystem requires selection and application of adapted experimental techniques. A short overview of the common techniques enabling characterization of smart fiber-reinforced composites is presented in the following sections. Stimulated by different levels of measurement, the described techniques have been differentiated into qualitative and quantitative ones.

The inspection techniques connected with the testing of electronic components and software procedures have been excluded from the following catalogue. Adequate procedures can be found, eg, in works of Gizopoulos (2006), Myers et al. (2012), or Beizer (1990).

7.2.1 Qualitative methods

The qualitative experimental methods deliver typically a visual representation of the investigated object. By the application of interpretation procedures—in elementary cases merely based on the experience of the operator—the characteristics of the examined object, eg, cracks, delamination, air pockets, or component misalignments can be extracted.

7.2.1.1 X-ray computed tomography

The X-ray computed tomography, commonly abbreviated as *X-ray CT*, is a nondestructive testing method used to analyze the internal structure of objects. During testing, the examined object is X-rayed from different angles and the resulting set of projections is reconstructed to a three-dimensional volume (Fig. 7.1). The X-ray CT technique is based on the differentiation of different materials by their dissimilar X-ray absorption characteristics. After the penetration of the sample, the rays exhibit the transmitted intensity I_1 that is defined by the intensity of radiation I_0, the absorber thickness h, and the attenuation coefficient μ according to the equation:

$$I_1 = I_0 \exp(-\mu h) \tag{7.1}$$

Since the attenuation coefficient depends on the atomic number and the density of the examined material, it is possible to differentiate between different materials (DIN Deutsches Institut für Normung e.V., 1997).

The spatial resolution of the current industrial X-ray CT systems is in micrometer range, which allows for identification of even single filament breakage in the fiber-reinforced composite structure. The X-ray CT has found wide application in materials testing and characterization (Bull et al., 2013). While the measurement capabilities of X-ray CT are very high, the attainable spatial resolution is limited by some difficulties on the stage of interpretation of results, which is caused by measurement noise and artefacts. Artifacts may have a different nature of occurrence, such as physically based (eg, X-ray beam hardening), scanner-based (eg, ring artifacts occurred during the reconstruction of measurement data to the three-dimensional array or the artifacts caused by a cone beam effect). The detailed description of possible artifacts occurred during CT scanning can be found in Barrett and Keat (2004).

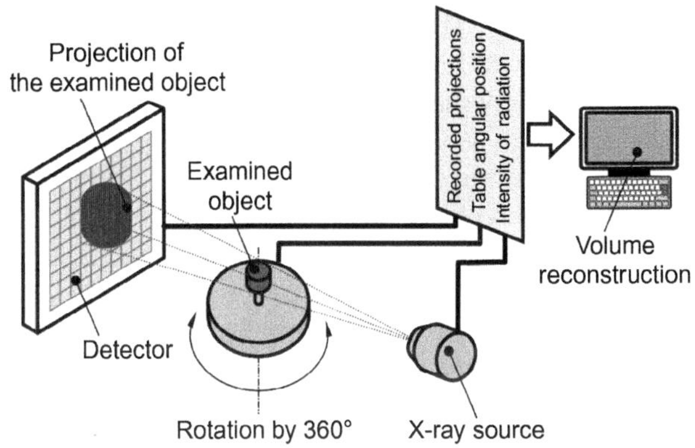

Fig. 7.1 Schematic overview of the operating principle of the X-ray computed tomography.

In light of characterization of smart composites, the X-ray CT can be used for the determination of breakage in the fibers, manufacturing related defects like air inclusions or resin-rich regions, soldering quality of the electronic components, positioning accuracy, mechanical caused damages, and geometrical set-up of actuator sensor systems. The limiting factor is the achievable spatial resolution resulting from the observed region as well as the total size of the analyzed component.

7.2.1.2 Micrograph analysis

The integration for example, of brittle piezoceramic components into fiber-reinforced composites requires extensive investigations with regard to the quality and failure analysis of the created smart structure. In this context, the micrograph analysis means a suitable method for the detection of microscopic interactions between the integrated piezoelectric components and the composite. Hereby, interfaces, transition zones, and failures in the composite (eg, delamination, fiber cracks or fiber-shifting) as well as in the functional element can be detected. For these investigations the region of interest has to be cut off the composite part. The resulting samples are embedded separately into a special thermoset embedding compound (eg, epoxy or acrylic resins). Afterward the samples are ground by a refining series of grinding surfaces (general grits: 500, 1200, 2400) by the utilization of a polishing machine (eg, System consisting of RotoPol-15, RotoForce-1, and Lupo dispenser, Struers GmbH). This guarantees a scratch free and plane surface. With regard to a flat and very reflective surface, which is necessary for micrographs with a high resolution, a subsequent polishing has to be performed. Therefore, a smooth polishing by an emulsion with dispersed diamond particles up to a grain size of minimum 3 μm (9 μm, 6 μm, 3 μm) and a cooling liquid is usually used. The emulsion and cooling liquid are dosed alternately to a polishing cloth and the samples are polished until the surface is completely free from scratches.

The prepared samples are investigated by a microscope with an attached digital camera. Different resolutions can be selected in accordance to the dimensions of the determined details. These are protocolled by a snapshot, which can be further processed by special image correlation or analyzing methods. In Fig. 7.2, exemplary microsections of material homogeneously integrated thermoplastic compatible piezoceramic modules (TPM) in fiber-reinforced thermoplastic composites are illustrated. Fig. 7.2A shows an ideal integration with a well-marked melting area of the TPM carrier film and the thermoplastic matrix of the composite. Compared with this, Fig. 7.2B shows an inaccurate integration with debonding effects and trapped air caused by low processing temperatures and pressures, which is accompanied by a barely affected TPM carrier film, which also causes only a small melting area between the composite matrix and the thermoplastic carrier film.

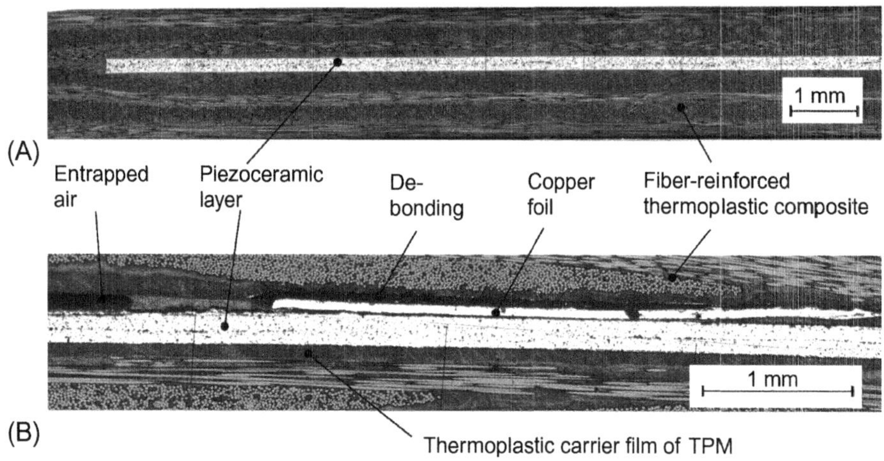

Fig. 7.2 Exemplary microsections of fiber-reinforced thermoplastic composites with integrated TPM: (A) ideal integration of a TPM; (B) integration with debonding effects and entrapped air.

7.2.1.3 Ultrasonic testing

A separate category of techniques for the qualitative description of smart components is based on an analysis of changes of ultrasound wave characteristic in terms of shape or amplitude after an interaction with the examined object. In particular, short-term, high-frequency waves are generated by a transducer and transmitted through a coupling medium, for example, gel, water, or air to a test object (Fig. 7.3). Several alternative measurement modes are possible (DIN Deutsches Institut für Normung e.V., 2010). A short selection of the most typical ones is summarized as follows:

Through transmission technique: An examination technique where the material is characterized based on the intensity of mechanical energy of the ultrasonic pulse after it has passed through the examined object.
Pulse echo technique: A technique in which the reflection of an ultrasonic pulse on material heterogeneities is analyzed in order to characterize their in-depth position and extension.
Time-of-flight technique: A technique where the relation between the pulse paths at various positions of examined object or wave angles of incidence are used for the description of the acoustic impedance heterogeneities.

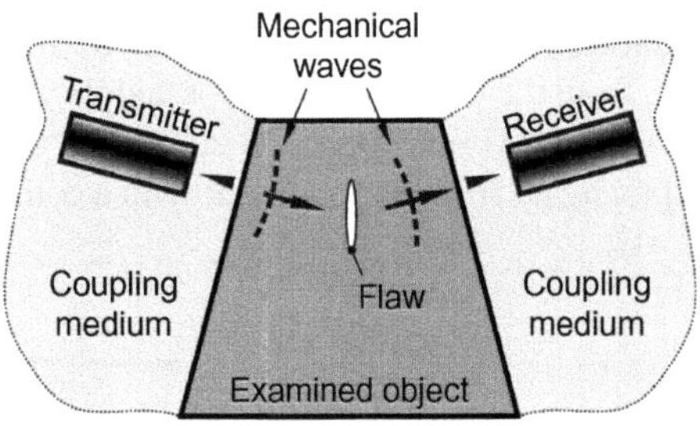

Fig. 7.3 Principal configuration of the system for the ultrasonic material testing.

Even though the ultrasonic testing has numerous advantages, such as the determination of the material stiffness and damping in wide frequency band and the delivery of accurate results even for dispersive materials, application of this technique for fiber-reinforced materials is connected with wave scatter. The wave scatter—resulting from the influence of the edges or from the intrinsic inhomogeneity of fiber-reinforced composites—causes problematic interpretation of the measured signals due to the presence of visual artifacts.

7.2.2 Quantitative characterization

Using quantitative methods, specific properties of smart composites are characterized. These include, in the case of sensors, the electric charge resulting from defined deformation of the integrated piezoelectric transducer or in the case of actuators, the resulting deformation of the composite structure caused by defined voltages applied to the transducer. In addition to the characterization of specific properties, quantitative methods can be used for quality assurance and for functional testing of smart composites.

7.2.2.1 Electric impedance

The aim of the integration of piezoelectric transducers is to implement sensory or actuatory properties. Both properties are mostly based on energy conversion. Energy converters frequently have the property of symmetry. That means the energy can flow in both directions and the energy flow on one side is influenced by the energy flow on the other side. A well-known example from the field of electrical engineering is the transformer. In principle, the model of the transformer is applicable to all symmetric energy converters. In the case of piezoceramic transducers, in particular also those which are embedded in mechanical structures, electrical energy is converted into mechanical energy. A pure piezoceramic disc converts electrical oscillations into mechanical vibrations, whose resonant characteristics depend on the material parameters and the geometry. The mechanical vibrations in turn affect current and voltage across the element. A simple equivalent electrical circuit diagram consisting of a parallel circuit of a capacitor C_p with a series resonant circuit (C, R, L) can be found in Fig. 7.4A. This circuit has a series resonance f_s (series resonant circuit) and a parallel resonance f_p (parallel capacitance C_p and inductor L). If the converter is embedded in a structure (eg, a composite) the impedance behavior changes since the ceramic is mechanically connected to the component. The structure influences both the resonances, as well as their peaks. That means, from the electrical behavior and its variation, conclusions on the overall system can be drawn. In the electrical equivalent circuit diagram, that can be displayed, for example, by an additional element with a complex behavior in series with the circuit (Fig. 7.4B).

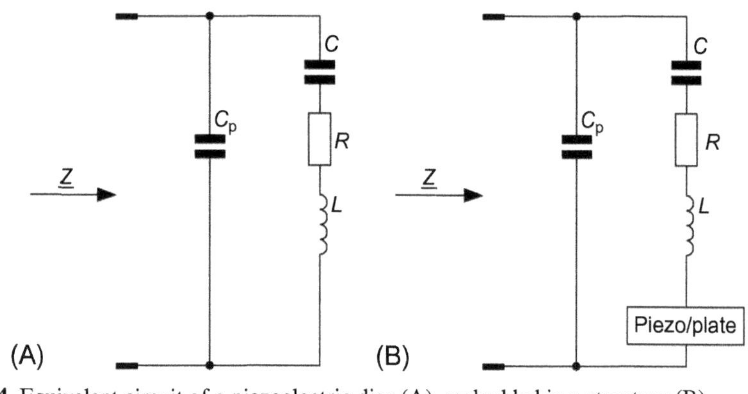

Fig. 7.4 Equivalent circuit of a piezoelectric disc (A), embedded in a structure (B).

To characterize the behavior of a piezoelectric disc it is helpful to use the complex impedance Z, ie, the ratio of voltage to current. By varying the exciting frequency, a typical frequency response curve of such elements can be obtained with a minimum amount of the impedance $|Z|$ in the case of series resonance and a maximum in the case of parallel resonance. The two frequencies also represent the respective proportion of energy forms, the mechanical and electrical energy (Fig. 7.5).

The ratio of both energy forms is defined as the coupling factor k_{eff}. It can be calculated from the two frequencies and is a measure for assessing the performance of such a transducer:

$$K_{\text{eff}}^2 = \frac{\text{Stored Mechanical Energy}}{\text{Total Stored Energy}} = \frac{f_p^2 - f_s^2}{f_p^2} \quad \text{(Ruschmeyer, 1995)} \tag{7.2}$$

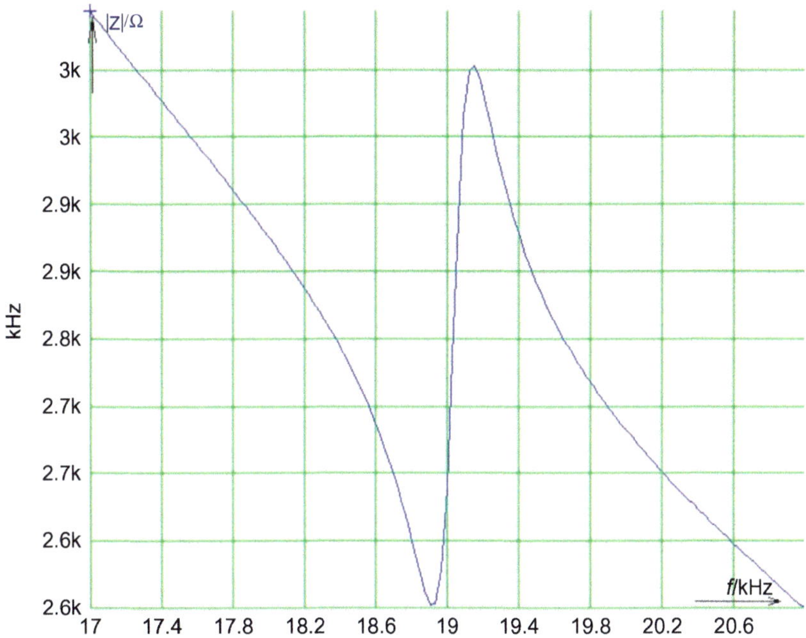

Fig. 7.5 Complex impedance versus frequency of a piezoceramic element.

7.2.2.2 Experimental modal analysis

In the past three decades, modal analysis has become a major technology in the quest for determining, improving, and optimizing dynamic characteristics of engineering structures (He and Fu, 2001). Modal analysis is used for the identification of inherent dynamic properties of a system, ie, natural frequencies, damping factors, and mode shapes. In general, the dynamic properties of the examined system are characterized based on the measurement of a vibrational answer of the system to a known or at least partially known excitation (Fig. 7.6). Herein, the excitation is typically realized using a modal exciter or impact hammer connected with force transducers as well as by application of other forms of excitation like chopped air stream, acoustic excitation, structure-integrated actuators or electromagnet soft iron anchor pairs. On the other hand, the vibrational answer can be determined using a great number of different transducer

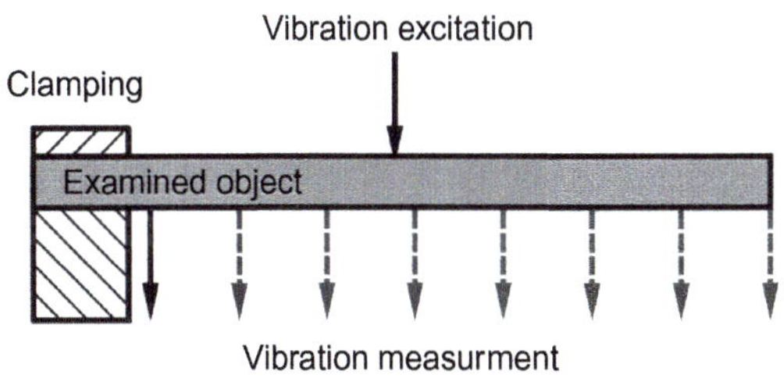

Fig. 7.6 Principal set-up for the identification of structure dynamic properties in roving sensor mode (grey arrows depict temporary positions of the vibration sensor).

types, eg, microphones, accelerometers, laser vibrometers, structure-integrated sensor networks, high-speed cameras, etc. The selection of proper combinations of excitation-measurement from the vast number of possible ones requires the expertise of the operator in order to guarantee reliable data.

The experimental modal analysis is a popular technique which can be used for the smart structure characterization; the experimental modal analysis is applied for the verification of active functions like active vibration damping or active shape morphing in order to influence the dynamical behavior. Additionally, the experimental modal analysis can be applied for the quality assurance or as a basis for structural health monitoring capabilities.

7.2.2.3 Cyclic three-point-bending

For measuring and quantifying the sensor characteristics of integrated piezoelectric transducers, different methods are available. In Section 7.2.2.1 it was shown that by resonant excitation of the structure and the then determined electrical properties, conclusions on electromechanical coupling and the associated efficiencies can be drawn. Another option is to deform the structure by mechanical loads and to measure the resulting electrically induced potentials or charges. This method also has the benefit that the test parameters can be modified and adjusted continuously online. It is also close to potential applications and future practical uses. The measuring principle is shown in Fig. 7.7.

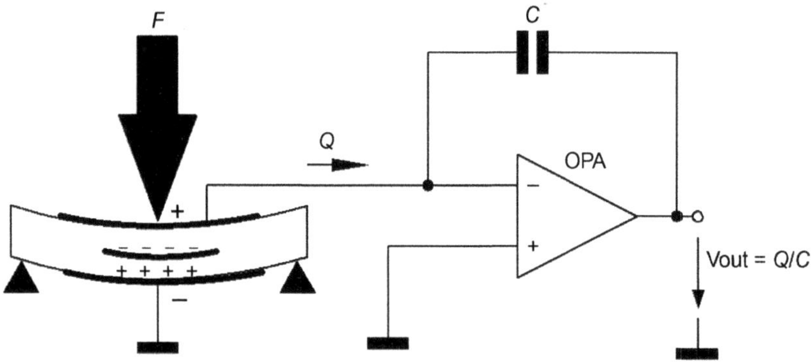

Fig. 7.7 Measuring principle of cyclic three-point-bending.

The arrangement of the composite sample with integrated piezoelectric transducer represents a bending bar. If this sample is deformed by a vertically applied force *F*, the integrated transducer is deformed, thus producing charges on the surface due to the piezoelectric effect. This in turn influences charges on two electrodes, positioned on the upper and lower side of the transducer.

Preferably, the charges are amplified by a charge amplifier according to Fig. 7.7 and converted into a voltage. The advantage over a simple voltage amplifier is the largely linear frequency response and the associated lower noise, which shows at least in a particular frequency range a high-pass character. The operational amplifier fulfills the task of a charge converter. Because of its finite electrical properties (input impedance, gain) the generated charges Q are degraded after a certain time. This charge loss must be compensated, which means the three-point bending must be cyclic.

7.2.2.4 Optical methods

Accurate measurements of 2-D and 3-D deformations of smart structures can be performed by contactless optical measuring methods based on the digital image correlation (DIC) principle. This starts by the measuring of the actuator performance based on a voltage-strain-hysteresis (Fig. 7.16) over the deformation measurement at harmonic vibration analyses to complex motion studies of compliant mechanisms. One method is the use of stochastic patterns to detect the deformation of the structure by measuring the changes in the pattern under loaded constitutions. In this case, comparable measuring systems like the ARAMIS system (GOM mbH) can be used, which enable the evaluation of 3D-surface coordinates, 3D deformations, and the plain strain tensor. Therefore the specimen has to be marked by a stochastic gray-value pattern. This is fragmented into facets (quadratic measuring fields) which are further divided into a predefined number of measuring points (see Fig. 7.8). The deformations of these facets are calculated by the transformation of the origin or undeformed to the deformed coordinate system of each facet (Sutton et al., 2009).

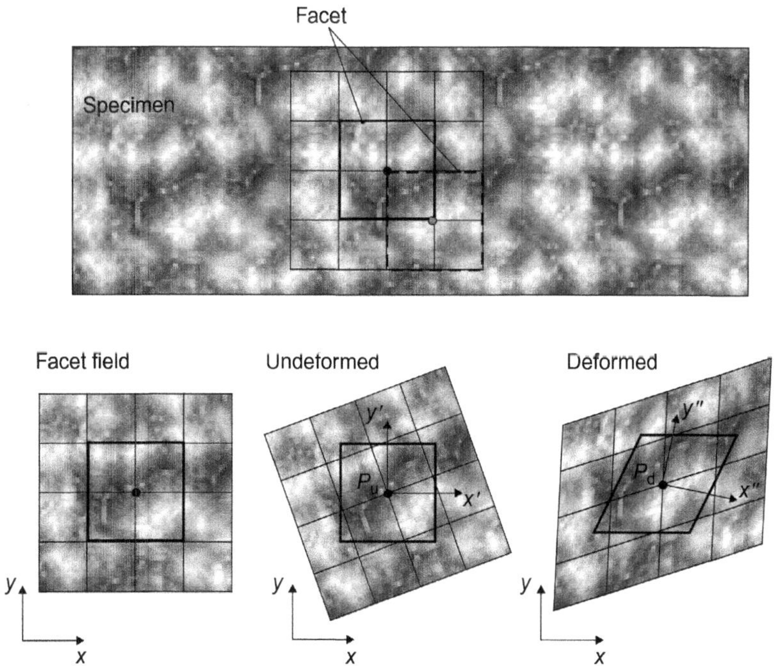

Fig. 7.8 Measuring principle using ARAMIS—fragmentation of the specimen in facets and detection of the facet and the coordinate system in a deformed state.

By tracking the facet shape, the deformation can be calculated by the tensor of the gradient of deformation F, the transformed line element $d\vec{x}$, and the origin line element in initial state $d\vec{X}$ using the following equation:

$$d\vec{x} = F \cdot d\vec{X} \tag{7.3}$$

Furthermore the tensor of the gradient of deformation is used for the determination of strain by using the tensor of rotation R and the tensor of the aspect ratio U

$$F = R \cdot U \tag{7.4}$$

whereby the tensor of the aspect ratio is defined as

$$U = \begin{pmatrix} U_{11} & U_{12} \\ U_{21} & U_{22} \end{pmatrix} = \begin{pmatrix} 1+\varepsilon_x & \varepsilon_{xy} \\ \varepsilon_{yx} & 1+\varepsilon_y \end{pmatrix} \tag{7.5}$$

A second method is the tracking of reflective points for the determination of displacements. Therefore, similar measuring systems like PONTOS (GOM mbH) can be used. The PONTOS measuring system as shown in Fig. 7.9 provides for the measuring points information about 3D coordinates and displacements, deformations, speed, and acceleration. Therefore, the investigated structure has to be defined by a reasonable number of markers. After calibration, the system records the position changes of these markers. These measurements can be processed by the system's software or directly pass-through for secondary software.

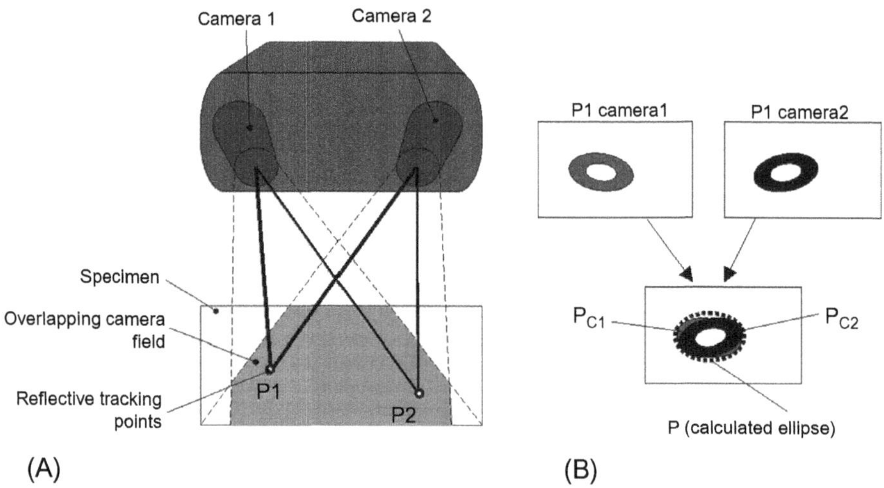

Fig. 7.9 PONTOS system: (A) Measuring set up and (B) detection of measuring points by ellipse calculation.

7.2.2.5 Alternative methods for the characterization of smart fiber-reinforced composites

Apart from the techniques presented in the previous sections, there are numerous approaches that can be applied for the characterization and testing of smart composites. The alternative methods briefly described in this section seem to have the potential to become *useful* for the characterization of smart fiber-reinforced composites.

The *in situ* computed tomography, which is a combination of X-ray CT and a universal testing machine, can be used to detect closed cracks (Holeczek et al., 2014). Under the influence of external loads the cracks open and can be visualized using the X-ray CT. Additionally, the *in operando* behavior, including the long-term stability of smart components, can be studied.

In infrared thermography the detection of adhesion or cohesion losses is possible based on the presence of "hot spots"—regions where an increased amount of energy is dissipated as heat (Krishnapillai et al., 2005; Wu and Busse, 1998). Herein, the thermographic cameras detect radiation in the infrared range of the electromagnetic spectrum and produce images of that radiation, called thermograms.

Another technique to detect the regions in which the mechanical energy is dissipated is the method based on tuning the interference of solid waves (Kostka et al., 2015). The method relies on the assumption that a standing wave—originating from the interference between two counter-propagating waves—can be considered as a weighting function to determine the distribution of dissipated energy. The dissipated energy in the regions with the maximal standing wave velocities (anti-nodes) has the highest weighting, whereas the weight of regions with zero oscillation deflection (nodes) is equal to zero. A shift of the node and antinode position, eg, by a modification of the phase of the travelling waves, causes a defined change of the weight distribution applied to the analyzed regions. By analysis of introduced work as a function of phase of the travelling waves, information regarding the damping distribution can be extracted.

The measurement of change of electric resistance or capacity of electrically conducting systems can deliver interesting insight into the quality of subsystems. The presence of cracks or "cold solder" joints causes fluctuation of the electrical properties under the mechanical load (Todoroki, 2003). Hence, by detection of such phenomena, the conclusions can be drawn regarding the current system state.

Time domain reflectometry (TDR) can be used to monitor relative displacements between structural elements (Su, 1998). In the TDR technique, an electromagnetic pulse is sent down the cable. If heterogeneity of wave impedance within the cable is present, a reflection of the electromagnetic pulse can be detected. Based on the time delay of this reflection, the position of heterogeneity can be calculated.

The electromagnetic emission (EME) can be used to investigate the material behavior, especially under highly dynamic loading conditions (Astanin et al., 2012). Based on the hypothesis that one of the sources of EME is fiber-matrix-debonding, a reliable methodology for the detection and classification of the complex failure phenomena in composites is constituted. Since different materials have varying relative dielectric constants, the formation of electrical double layers takes place at the boundary of fiber and matrix. During fiber matrix debonding, the layers of different dielectric constants within the electrical double layer zones are moving apart. As a consequence, areas with a concentrated electric charge are formed. With the lapse of time, the discharge and a related deconcentration over the material volume occurs.

Another technique to characterize the smart composites is constituted on fusion and processing of data from different sources (Katunin et al., 2015). Considering the high sensitivity to the singularities the discrete wavelet transform (DWT) is one of the best techniques used for feature detection and identification.

7.3 Case studies

In the following, case studies using selected characterization techniques are presented. Hereby, both sensor and actuator properties as well as potential applications are considered and specific results are illustrated. Beyond the presented studies, the measurement principles can be adopted for other applications, using different transducers and structures.

7.3.1 Cyclic three-point-bending of long fiber-reinforced polyurethane composite structures with novel integrated sensors

The production of prefabricated piezoceramic modules and their integration into active composite structures are often costly and time consuming and represent separate production steps. Therefore, a technology for process-integrated manufacturing and embedding of novel sensor modules has been developed (Weder et al., 2012). Based on the long fiber injection spray coat method, this technology is characterized by processing of reactive polyurethane systems and glass fibers, and especially moderate processing conditions in terms of pressure and temperature (Geller et al., 2015). In addition, the spraying process allows gentle embedding of the functional elements and even the electronic evaluation systems. In order to proof the functionality of the integrated sensors, composite samples are characterized using cyclic three-point-bending tests.

7.3.1.1 Materials and manufacturing

As starting components, free-flowing piezoceramic semifinished products like pearls or short fibers based on lead zirconate titanate (PZT) were used. During manufacture, these were placed on a lower electrode, forming a piezoelectric functional layer. By depositing the upper electrode on the functional layer, the sensor assembly was finalized, see Fig. 7.10A. For completely embedding the sensor in the composite material, the sensor assembly was impregnated during the consolidation by the expanding matrix in thickness direction (Fig. 7.10B). Therefore, porous electrode structures such as metal-wire meshes were used.

For the characterizations presented, sensor specifications with parameters shown in Table 7.1 have been used.

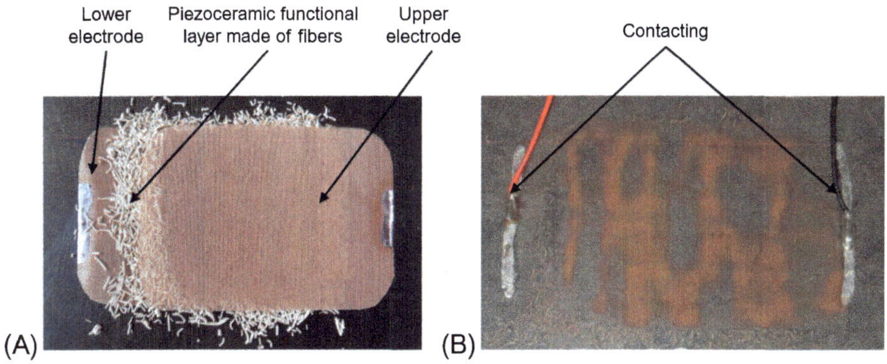

Fig. 7.10 Sensor module with fibers and porous electrodes before (A) and after embedding (B).

Table 7.1 Specifications of used piezoceramic components and electrodes

Piezoceramic components	Pearls	Fibers
Diameter (mm)	1.5	0.5–0.8
Length (mm)		1–10

Electrodes	Type E1	Type E2
Material	CuSn6	
Mesh size (mm)	0.1	0.04
Wire diameter (mm)	0.065	0.032

7.3.1.2 Experimental

To proof the functionality of the newly developed sensor modules, they have been characterized using dynamic three-point bending as described in Section 7.2.2.3. In Fig. 7.11, the corresponding test setup is shown. For these characterizations, sensors in four specifications using fibers (F) and pearls (P), and both types of electrodes (E1 and E2) were integrated into glass fiber-reinforced polyurethane composites. After integration, the sensors were contacted and polarized. When impregnating the sensor assembly during the manufacturing process, the formation of polymeric bilayers between piezoelectric functional layers and electrodes may result due to the expansion process of the polyurethane matrix. These layers affect the polarization behavior and the resulting sensor properties. Therefore, the influence of such polymeric bilayers on the polarization behavior was analyzed (Geller et al., 2014). Due to the low permittivity of the polymer matrix, relatively high voltages are required for polarization. The integrated sensor elements were tested before and after polarization, whereby polarization voltages of 2.5, 3.3, 4.1 and 4.9 kV were examined.

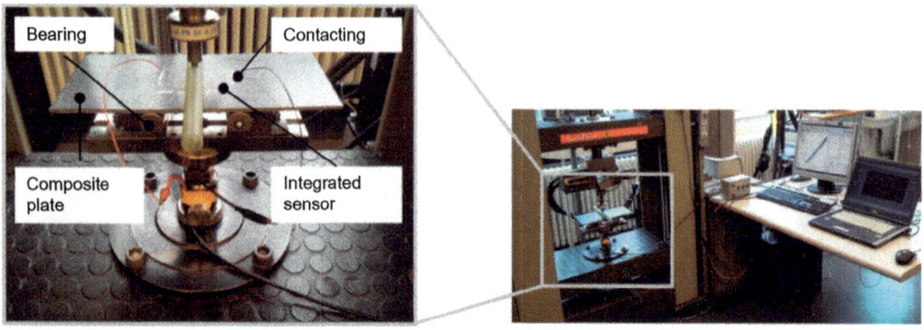

Fig. 7.11 Test setup for cyclic three-point-bending of integrated sensor elements.

7.3.1.3 *Results*

The results show that sensors with a functional layer of fibers generate higher electric charges compared to sensors with pearls (Fig. 7.12). This can be attributed to the greater deformation of the fibers during bending which resulted in larger displacement of the positive and negative charge concentration within the piezoceramic. An influence of electrode specification could not be detected.

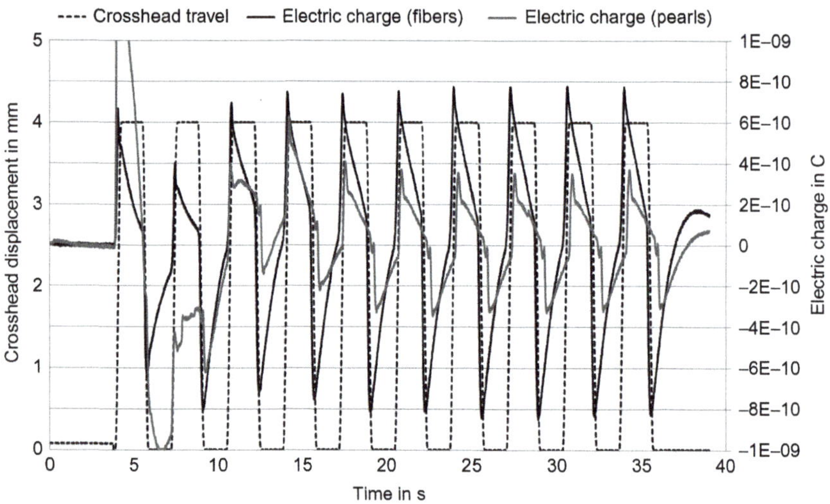

Fig. 7.12 Detected electric charge during cyclic three-point-bending in dependency of crosshead displacement.

As expected, with increasing polarization voltage, higher electric charges were generated. In Fig. 7.13 the detected charges of an integrated sensor with electrode type E2 and functional layer made of fibers are shown in dependency of the applied polarization voltage. Even if the values of the determined electric charges were relatively small, the functional capability of the novel sensors could be proven.

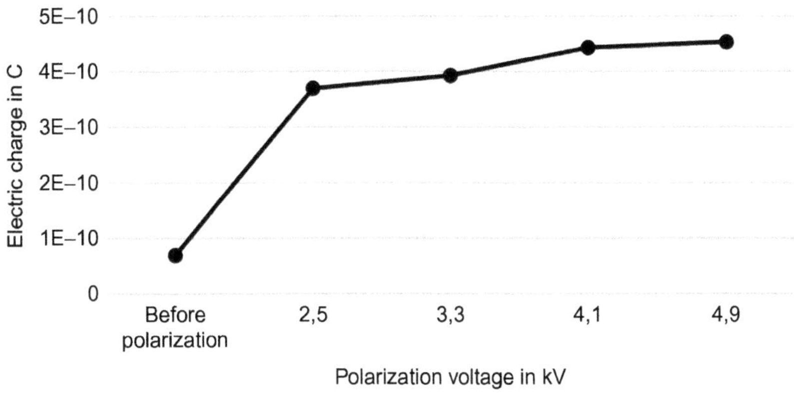

Fig. 7.13 Electric charge in dependency of applied polarization voltage.

7.3.2 Textile-reinforced thermoplastic structure with material homogeneous embedded piezoceramic modules

Fiber-reinforced thermoplastic composites show a high potential for series capable manufacturing because of the realizable short processing times (Neitzel et al., 2014). Due to the layered build up of the composite structures, additional functional elements can be integrated between the single layers, even during the manufacturing of the composite parts. Furthermore, the use of identic materials for the component matrix and carrier films for the functional elements allows for material homogeneous embedding of these elements. By implementation of additional functionalities, these parts enable inline measuring or actuation abilities, so that a time and cost intensive adhesive bonding of comparable active elements can be omitted. For a generic demonstrator structure, selected characterization methods are presented. On one hand there is the quality control by the use of destructive and nondestructive test methods, and on the other hand there is the performance measuring by means of DIC.

7.3.2.1 Materials, Components and manufacturing

The generic demonstrator shown in Fig. 7.15B consists of a glass fiber-reinforced polyamide 6 (GF-PA 6) component. Additionally, especially developed matrix identical thermoplastic compatible piezoceramic modules (TPM) were embedded in the demonstrator structure during its manufacturing process. In Fig. 7.14 the built up of different TPM configurations are shown, consisting of thermoplastic carrier films, metallized with electrode structures and piezoelectric functional layers (eg, piezoceramic plates, fiber composites, or printing pastes) in the middle.

For the structural component, semifinished fiber-reinforced PA 6 plates (type TEPEX®102-RG600(x)/47%, Bond Laminates GmbH) are used. By means of a hot pressing process, these plates were heated up over melting temperature, the TPM was inserted in the lay-up, and during the consolidation of the part under temperature and pressure, the TPM carrier film melted to the composite matrix. Consequently, the piezoceramic functional layers were material homogeneously and directly integrated

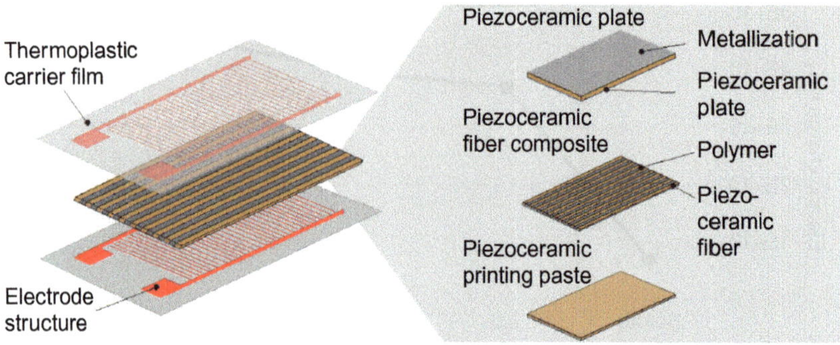

Fig. 7.14 Illustration of possible TPM built ups.

into the composite part, so that actuating and sensing functionalities are realizable with the generic demonstrator structure. In the following, the actuation of the embedded TPM and its resulting effect of the resonance behavior of the investigated structure are shown.

7.3.2.2 Test setup

The experimental validation of the actuation of material embedded TPM was performed by a resonance analysis using the PONTOS-Measuring system (GOM GmbH). For this measurement the structure was treated with special reflective markers and hung up using thin light ropes (see Fig. 7.15).

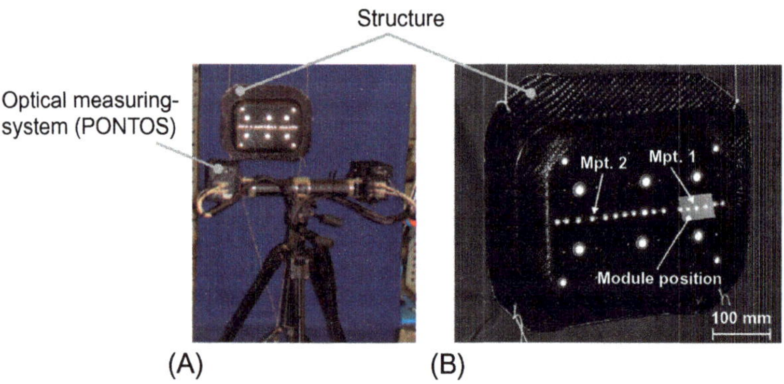

Fig. 7.15 Fiber-reinforced thermoplastic composite structure with reflective points, measuring points (Mpt.) and the module position.

The TPM is driven by a sinusoidal excitation using voltages from 0 to 150 V which is generated by a high voltage power amplifier. The analysis is performed in a frequency range of 10–350 Hz. Due to module position, an excitation within the second eigenfrequency of the structure at 287 Hz shows significant deflections. Using the PONTOS system, the out-of-plane displacement could be measured, which is one of the design criteria for structural components.

7.3.2.3 Experimental and quality analysis

For the quality control of the TPM itself, the stress–strain behavior of the modules was measured with help of the grey scale correlation procedure (ARAMIS). Fig. 7.16 shows the test setup including function generator, amplifier and DIC-System (5M-ARAMIS-system). For this quantitative measuring of material parameters the method of statistic evaluation within the DIC-Software was used. The arithmetic mean of the strain field was used as a rating characteristic of the module. In those kinds of measuring setups, there are up to 8000 data points over the active field.

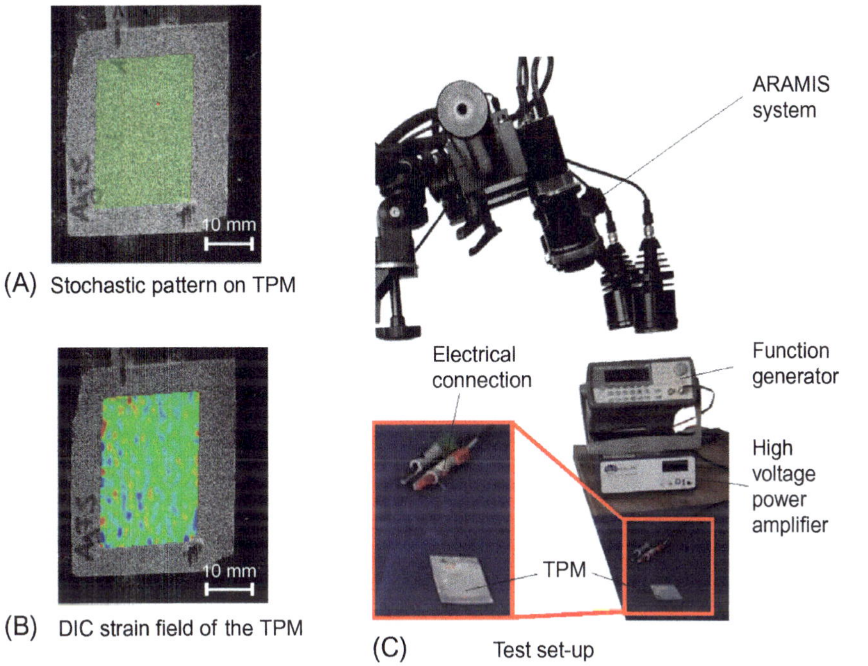

Fig. 7.16 Setup for module performance measurement.

Fig. 7.17 shows the measured strain hysteresis of TPM with thermoplastic carrier films made of polyetheretherketone (PEEK) and polyamide 6 in comparison to a conventional MFC actuator. The TPM was actuated utilizing a triangle signal and a frequency of 0.1 Hz. The free strain ε_f in the main actuation direction is displayed as a function of both the applied operation voltage U and the maximum electric field strength that arises over the module cross section.

Due to the rough application conditions of the module within the manufacturing process, an evaluation of the embedded module quality becomes necessary. On one hand, the active layer should be embedded without cracks, and on the other hand, the interface between the TPM and the structural material should be in an optimal condition. To detect cracks within the functional layer, an X-ray examination was used,

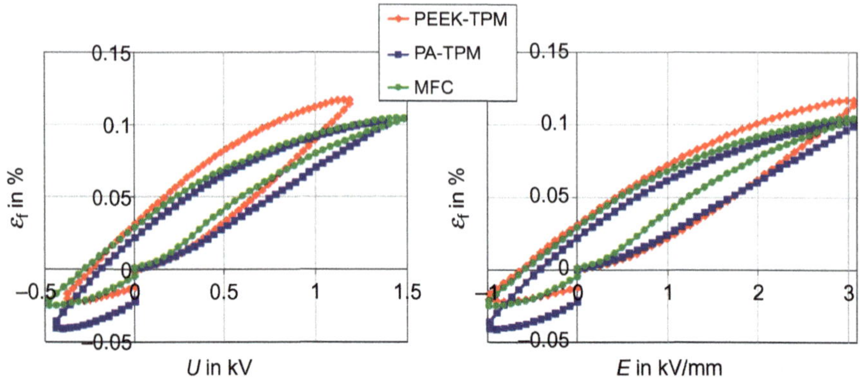

Fig. 7.17 Voltage-strain-hysteresis of different modules.

shown in Fig. 7.18A. Because of the material behavior of the functional layer (lead zirconate titanate) there is no possibility to characterize the interface between the module and the structural material by nondestructive methods like X-ray CT. Therefore, there is a need for destructive test methods like micrographs. In Fig. 7.18B a micrograph of the embedded module is shown. The investigation shows a homogeneous interface area without voids and delamination.

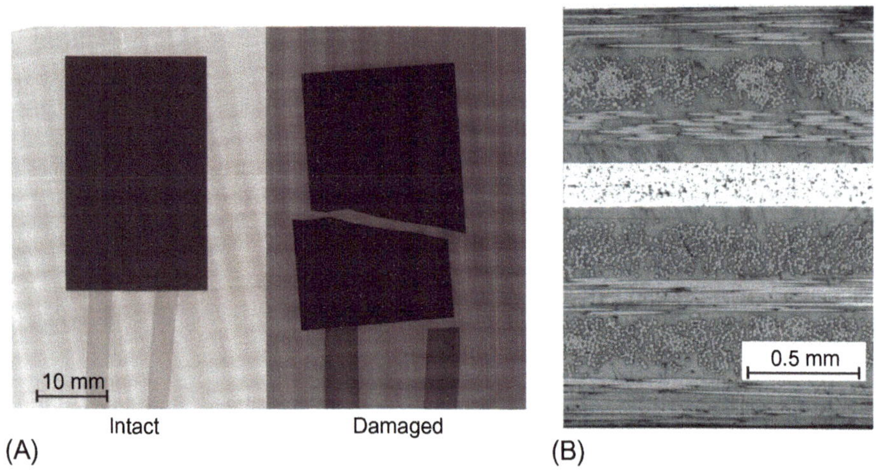

Fig. 7.18 Analysis of the integrated modules: (A) micrograph, (B) X-ray.

7.3.2.4 Results

For the measuring of the active thermoplastic demonstrator with frequencies up to 350 Hz, the DIC-System PONTOS HS was used in a special frequency dependent mode. Due to the need of system resolution with 1280×1024 pixel and in comparison of the maximum frame sampling rate (500 fps) the structure is measured by undersampling using aliasing. Fig. 7.19 shows the out-of-plane displacement of two points of the demonstrator structure at a sinusoidal excitation with 287 Hz. The deflection of the measured points is nearly 180 degrees out of phase so the structure is driven in resonant frequency. The diagraph also shows the phase shift between excitation voltage maximum and displacement maximum, which is reasoned in material and structural damping effects.

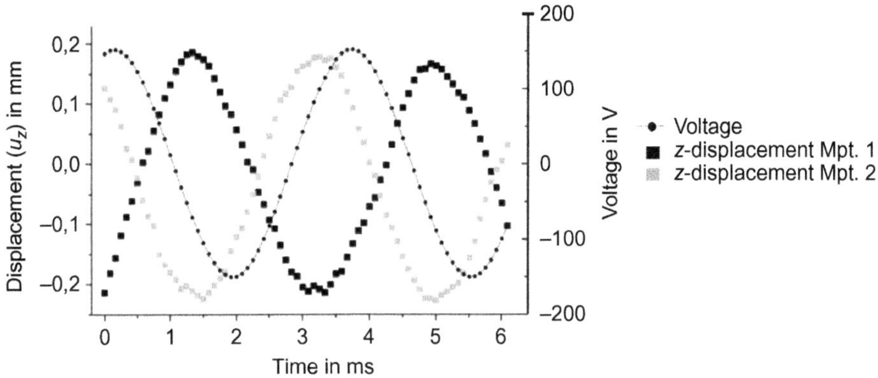

Fig. 7.19 Structural response of the demonstrator structure at 287 Hz.

7.3.3 Disc-shaped rotor with integrated active damping system

Every in-service structure is to a greater or lesser extent exposed to vibrations. These can be either structure born—for example, resulting from realized process characteristics—or foundation-borne like wind, noise, etc. The vibrations can, in the worst case, lead to the structure's catastrophic failure due to eg, fatigue or dynamic overload. An increase of vibration-caused issues can arise from the application of materials with relatively low damping properties like high-performance lightweight polymer-based composites. The application of material-integrated active damping systems suppress the vibrations by generation of vibration-counteracting forces using embedded elements. An application of such system is presented on example of a disc-shaped segmented rotor made of glass-fiber-reinforced polyamide and material-integrated sensing/actuating elements.

7.3.3.1 Experimental setup

For the assessment of the active damping function performance, the segmented rotor structure has been investigated under idealized conditions, ie, fixed clamping of the rotor on its inner radius, lack of external noise in the form of wind or foundation-borne vibrations. One of the rotor segments was excited to vibrations with the first eigenmode at 53.6 Hz using an externally placed electromagnet and rotor bonded soft-iron anchor (Fig. 7.20).

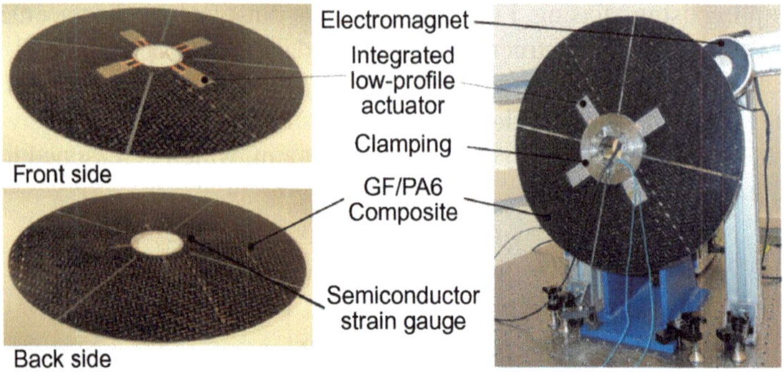

Fig. 7.20 Test stand for the damping performance investigation of the fiber-reinforced rotor with integrated active damping systems.

7.3.3.2 Results

For the active damping, a novel modified phase-locked loop (mPLL) algorithm was used. In the mPLL strategy the measured harmonic input signal was used for the generation of a synthetic harmonic signal with matching to the measured signal initial phase. This synthetic signal was then used as the control signal for the actuating elements. In comparison with typical active vibration control strategies (IMSC or MIMSC), such an approach is on one hand significantly more robust regarding noise sensitivity, and on the other hand offering simple online phase correction. Moreover, no information regarding modal parameters—as in typical control strategies—is necessary, resulting in significant simplification of the design phase.

The strain signal was measured using an integrated semiconductor strain gauge and transmitted after discretization by wire to the control unit with implemented mPLL control algorithm. Herein, the calculated suitable control signal was sent after amplification to the material-integrated low-profile actuating elements. Through the influence of the generated forces, the actual damping of the structure at this eigenform increased significantly, which resulted in reduction of the vibration amplitude. In Fig. 7.21, at the time point $t=3$ s the active damping is activated.

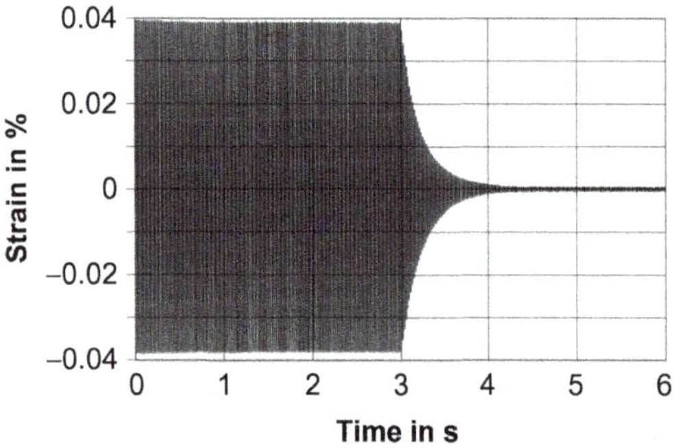

Fig. 7.21 Damping performance of the integrated system. The integrated damping system has been activated in the 3rd second.

It is clearly observable that the vibration has been reduced by approximately 98% in comparison with the passive system. Hence it is theoretically possible to design components that can operate in conditions which were formerly not possible.

7.4 Summary/conclusions

Textile reinforcements can be used for the generation of active fiber-reinforced composite structures. With regard to the presented investigations, the composite parts are functionalized by the integration of piezoelectric elements. These parts obtain electrical and electromechanical functionalities in addition to their excellent mechanical properties. The presented characterization and testing methods are useful for the structural and functional validation of textile-reinforced composites. They allow for a process-accompanied quality control as well as for the support of the developing process of novel function-integrative lightweight structures. Furthermore, the presented case studies demonstrate the application of specific methods for selected applications. The manufacturing process development of the presented active composite structures was accompanied by destructive and nondestructive testing methods in regard to the validation of the structural integrity of sensor/actuator elements and composite structure. Using the DIC, LDV, and electrical impedance analysis, the functional behavior of the presented active parts is approved and quantified. Consequently the presented chapter gives an overview of possible characterization and testing methods in regard to active composite parts with embedded piezoelectric elements.

Acknowledgment

This research is supported by the Deutsche Forschungsgemeinschaft (DFG) in the context of the Collaborative Research Centre/Transregio 39 PT-PIESA, subprojects A05, B04 and B06.

References

Astanin, V.V., et al., 2012. Characterising failure in textile-reinforced thermoplastic composites by electromagnetic emission measurements under medium and high velocity impact loading. Int. J. Impact Eng. 49, 22–30.

Barrett, J.F., Keat, N., 2004. Artifacts in CT: recognition and avoidance. Radiographics 24 (6), 1679–1691.

Beizer, B., 1990. Software Testing Techniques, second ed. Van Nostrand Reinhold, New York.

Bull, D.J., Sinclair, I., Spearing, S.M., 2013. Partial volume correction for approximating crack opening displacements in CFRP material obtained from micro-focus X-ray CT scans. Compos. Sci. Technol. 81, 9–16.

DIN Deutsches Institut für Normung e.V., 1997. DIN EN 1330-3: Non-destructive testing—Terminology—Part 3: Terms used in industrial radlographic testing. Trilingual version EN 1330-3: 1997, Beuth Verlag GmbH, Berlin.

DIN Deutsches Institut für Normung e.V., 2010. DIN EN 1330-4: Non-destructive testing—Terminology—Part 4: Terms used in ultrasonic testing. Trilingual version EN 1330-4: 2010, Beuth Verlag GmbH, Berlin.

Edery-Azulay, L., Abramovich, H., 2006. Active damping of piezo-composite beams. Compos. Struct. 74, 458–466.

Geller, S., Neumeister, P., Gude, M., Tyczynski, T., 2014. Studies on the polarisation behaviour of novel piezoelectric sensor modules. Sensors Actuat. A: Phys. 218, 162–166.

Geller, S., Winkler, A., Gude, M., 2015. Investigations on the structural integrity and functional capability of embedded piezoelectric modules. In: Proceedings of the 6th International Conference on Emerging Technologies in Non-Destructive Testing, Brussels.

Gizopoulos, D. (Ed.), 2006. Advances in Electronic Testing. Springer, Dordrecht.

He, J., Fu, Z.-F., 2001. Modal Analysis. Butterworth-Heinemann, Oxford, Boston.

Holeczek, K., Kostka, P., Modler, N., 2014. Dry friction contribution to damage-caused increase of damping in fiber-reinforced polymer-based composites. Adv. Eng. Mater. 16 (10), 1284–1292.

Hufenbach, W., Gude, M., Heber, T., 2009. Design and testing of novel piezoceramic modules for adaptive thermoplastic composite structures. Smart Mater. Struct. 18 (4), 1–7.

Katunin, A., Dańczak, M., Kostka, P., 2015. Automated identification and classification of internal defects in composite structures using computed tomography and 3D wavelet analysis. Arch. Civil Mech. Eng. 15 (2), 436–448.

Konka, H.P., Wahab, M.A., Lian, K., 2012. The effects of embedded piezoelectric fiber composite sensors on the structural integrity of glass-fiber-epoxy composite laminat. Smart Mater. Struct. 21 (1), 1–9, 015016.

Kostka, P., Holeczek, K., Hufenbach, W., 2015. A new methodology for the determination of material damping distribution based on tuning the interference of solid waves. Eng. Struct. 83, 1–6.

Krishnapillai, M., et al., 2005. Thermography as a tool for damage assessment. Compos. Struct. 67 (2), 149–155.

Lin, M., Chang, F.-K., 2002. The manufacture of composite structures with a builtin network of piezoceramics. Compos. Sci. Technol. 62 (7–8), 919–939.

Melnykowycz, M.M., et al., 2005. Integration of active fiber composite (AFC) sensors/actuators into glass/epoxy laminates. In: Proceedings of the SPIE 5761, 221–232.

Myers, G.J., Badgett, T., Sandler, C., 2012. The Art of Software Testing, third ed. Wiley, Hoboken, NJ.

Neitzel, M., Mitschang, P., Breuer, U., 2014. Handbuch Verbundwerkstoffe—Werkstoffe, Verarbeitung, Anwendung, second ed. Carl Hanser Verlag, München.

Ruschmeyer, K., 1995. Piezokeramik-Grundlagen, Werkstoffe, Applikationen. expert, Renningen-Malmsheim, 31.

Su, M.B., 1998. TDR monitoring systems for the integrity of infrastructures. In: Smart Structures and Materials 1998: Smart Systems for Bridges, Structures, and Highways. San Diego, CA.

Sutton, M.A., Orteu, J.-J., Schreier, H.W., 2009. Image Correlation for Shape, Motion and Deformation Measurements/Basic Concepts, Theory and Applications. Springer-Verlag GmbH, New York.

Todoroki, A., 2003. High performance estimations of delamination of graphite/epoxy laminates with electric resistance change method. Compos. Sci. Technol. 63 (13), 1911–1920.

Weder, A., et al., 2012. A novel technology for the high-volume production of intelligent composite structures with integrated piezoceramic sensors and electronic components. Sensors Actuators A 202, 106–110.

Wu, D., Busse, G., 1998. Lock-in thermography for nondestructive evaluation of materials. Revue Generale de Thermique 37 (8), 693–703.

Environmental textiles: Testing and certification

8

M.D. Teli
Institute of Chemical Technology, Mumbai, India

8.1 Introduction

The consumer-driven market of today demands performance textiles which can meet international quality standards and exhibit high performance with respect to the specific application's needs. For example, sportswear requires, in addition to a high level of comfort for the wearer, stretchability, moisture management, and temperature adaptability for protection from adverse weather. In cases of military textiles, additionally the material should offer good protection from hostile weather conditions. Some of the specific protective gear such as bullet-proof vests, require the material to be resistant to high impact. In the case of firefighters' clothing, in addition to the light weight and comfort, durable flame retardance and thermal resistance are also of great importance. For medical textiles, depending upon the final use of the products, the antibacterial property, biodegradability, in some cases superabsorbency and ability to provide sterile surroundings, etc. are of prime importance. A lot of conductive materials are used in wearable electronics and the issues relating to wash durability of such performance and wearability become quite significant. While in general when our surroundings are becoming increasingly noisy, the textiles offering sound absorbing property are of great value and hence such materials as well as smart fiber reinforced composites will have wide applications. These composites can be made of high performance oriented materials offering replacement of metal composites with enhanced performance per weight.

Environmental textiles within the framework of high performance textiles transcend from one category of speciality textiles to another. Many of these high performance textile materials are made of commonly known fibers and polymers using very similar processes and technologies. Hence, depending upon the application of the products in a particular category, they are classified, for example, military textiles, medical textiles, sports textiles, etc. The effect of these products on the environment during their manufacture and fabrication as well as during their service life and disposal will be mainly governed by the basic factors such as fibers, polymers, processes, and technologies involved in their making, and effect on environment. These main components of the product manufacture, in addition to the depletion of precious resources such as raw materials, water, and energy, decide the impact on water pollution, air pollution (Green House Gases), noise creation, etc. Many of these modern technologies are considered pivotal in exhibiting high performance; however, they are not without risks and limitations; for example, many of the functional finishes,

Performance Testing of Textiles. http://dx.doi.org/10.1016/B978-0-08-100570-5.00009-8

especially those applied on textile materials for depicting antibacterial property, UV protective property, self-cleaning ability, or super hydrophobicity are at their best performance when used as nanomaterials. However, such finishes containing nanomaterials may not be completely safe for use as next-to-skin garments. There was always a risk of such nanoparticles finding an entry into our body through the pores of the skin and thus interfering with human health and safety. A number of researchers are studying the health risk of such high performance materials containing nanoparticles. In other words, when environmental textile products are to be considered for testing and application, one has to be doubly sure that these tests are done keeping in view the safer and tolerance limits of the various intermediates, chemicals, colorants, and heavy metals used in it. The environmental testing of textiles has to be seen through the lens of the raw materials used, as well as the manufacturing process. It is well established that the asbestos fiber is carcinogenic, and although it is known as a flame-retardant material with a very high limiting oxygen index, its carcinogenicity creates an alarming risk to human life, so irrespective of testing standards and the performance analysis, such a material is simply banned because of its failure to fulfill the basic necessity of being safe for humans. Thus, when environmental textiles which span through various categories of high performance materials are studied, the underlying expectation from all of them is that they are supposed to exert the least environmental impact and offer maximum human safety. The testing methods of these materials to assess their impact on the environment and human life should take into account the gaseous emissions, effluent load, biodegradability, energy consumption, and health and safety of the workers in the manufacturing unit and the general consumers of such products. The life cycle analysis, which takes into account the impact during their use and disposal, is also equally important for determining the carbon footprint of such products.

8.2 Historical background

Conventional textile processing is notorious for creating environment pollution—both air and water pollution, due to very high levels of unused dyes, pigments, and chemicals in the effluent. Naturally any laxity in treating the effluent leads to environmental degradation and adds to the miseries of people living in the vicinity. During the 1990s, developed countries realized that it is extremely difficult to make such operations sustainable with the available technology, and hence decided to shift such operations into so-called third world countries. Increasing labor costs and stringent environmental laws in the western developed world coupled with pressure exerted by consumers' rights organizations brought to a halt such operations in developed countries. The natural effect of this self-imposed development was reflected in an enhanced demand for such products from South East Asian countries, which considered it a great boon for their textile industry, enabling them to earn valuable foreign exchange. In the initial phase, they too did not worry much about environmental laws, and unbridled manufacturing activities and supply to developed nations certainly boosted their economy. Over the period, however, ill effects of basic flaws in technology and way the textiles were processed became increasingly apparent even in these countries. Their environmental protection lobby became strong, people working in nearby textile units and workforces were facing unsavory conditions, and consumers also started becoming increasingly aware. The importing countries in the developed world also realized this and started putting pressure on the suppliers with respect to the quality standards of textiles and garments imported from their country. Germany was among those on the forefront of this movement.

8.3 German ban on azo dyes

Some of the dyes of direct and acid classes based on azo chromophore used for coloration of textiles are prone to reduction due to the reductase class of enzymes found in human perspiration. Once the azo chromophore is reduced, the color fades or decomposes, and some of the reduced products which are amines have carcinogenicity. These amines thus potentially pose a danger to human health and also to aquatic life in the environment. Many times in developing countries, when manufacturing activities using such dyes were at a peak, a number of reports indicated that textile effluents containing such azo dyes created great damage to the ecosystem. Knowing the harmful effect of such azo dyes, the importers from western countries set a ban on the use of such dyes, and in 2002, Germany and European Union countries banned azo dyes, which have the potential to break down into one of any of the 24 carcinogenic amines (refer to Fig. 8.1).

The German azo ban came into existence in 1994 due to the amendment of the German consumer goods ordinance to ban azo dyes which have the potential to release 20 carcinogenic amines on reductive cleavage of azo bonds. Over the years, the metabolic cleavage of azo dyes by bacteria found in the digestive tract of humans or by enzymes has been proved. In 1999, further amendments to the German ordinance included four more amines, mainly orthoanisidene, *para* amino azo benzene, and two xylidene isomers (DyStar Customer Service, 2000). However, there are a number of azo dyes of acid and reactive classes such as Acid Red 337, CI Acid Black 60, CI Reactive Orange 13, C I Reactive Red 84, etc. which are presently being used in the industry, since on their reduction the amines released do not pose any danger to the human life.

During the initial years when economies of many countries were liberalizing and winds of globalization had just started blowing, long before 2002, many of these developing countries, including India, Bangladesh, Sri Lanka, etc., did not take the German ban on Azo dyes seriously and considered it a trade barrier. However, when the German ban (which was more well-known that the EU ban) was implemented more effectively in 2002, those countries depending on valuable foreign exchange by exporting to Germany and EU countries had to immediately substitute the azo dyes in their range with safer dyes. The EU legislation in section 43 of annex XVII of REACH indicates that the dyes capable of reducing the carcinogenic aromatic amines in more than 30 ppm in textile articles may not be used. However, the azo pigments, being insoluble, are not restricted by this ban (Puntener and Page, 2004).

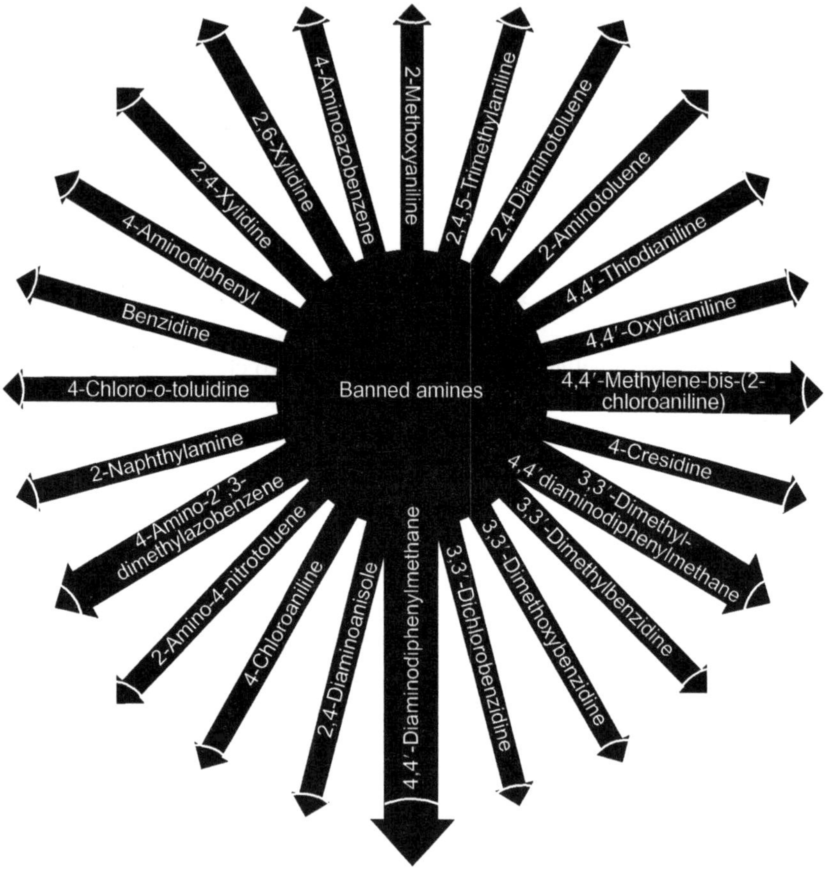

Fig. 8.1 List of 24 banned amines.

The detection of such banned azo colorants used on cellulose and protein fiber is done as per the Method B 82.02-2 and for polyester Method B82.02-4 (Methods for Determination of Certain Aromatic Amines Derived From AZO Colorants—Part 1: Detection of the Use of Certain AZO Colorants Accessible With and Without Extracting the Fibres) and they are legally recognized in the consumer goods' ordinance. The procedure indicates the textile sample be first treated with a citrate buffer solution for half an hour at 70°C, then sodium dithionate solution is added and the sample is stored for 30 min; further, after cooling, it is eluted with tertiary butyl methyl ether through silica column and then it is further concentrated and dissolved in methanol and analyzed for amines using thin layer chromatography, high performance liquid chromatography (HPLC), gas capillary chromatography (GC), and capillary electrophoresis. The quantitative analysis is done using HPLC/DAD (diode array detector or GC-MS). The country to country adoption of regulation in manufacture and use of these dyes varied by time scale. It is also dependant on types of articles used. A number of azo dyestuff manufacturers provide certification that they do not manufacture similar dyestuffs. In the initial list, there were 112 dyestuffs of direct and acid class which are banned as per the German consumer ordinance, and there are another 30 dyestuffs, though yet not featured on the banned list, can release the carcinogenic amines (Hunger and Sewekow, 2003).

Textile effluents, which are the most polluting, not only in terms of volume but also consumption, pose the threat of great damage to the environment and human health. One of the estimates indicates that 200,000 tons of dyestuff is released into the environment annually. The concentration of azo dye sometimes is as high as 500 ppm (Chequer et al., 2013). Many developed countries have followed the EU model and prescribed limits, and many times, even in developing countries, there are limitations enforced in the total dissolved solids of the effluent to be discharged after treatment. In 1992, aromatic amines causing cancer, particularly benzidine and 2-naphthyl amine, were established (LGC, 1999; Chequer et al., 2011; Puvaneswari et al., 2006). The impact of azo dyes released in the effluent in Brazil has also been reported in terms of mutagenic activity seen in one of the rivers (Lima et al., 2007). A number of different methods for remediation of water has been recommended, and effluent treatment plants became essential in various countries. The adverse effect on the environment of untreated effluent going into the river beds has not only done irreversible damage to the environment, but also to human health, and many people living nearby such industrial sites have been the victims of such activities. Under the World Trade Organization, the importing country sets forth the quality parameters of the goods imported into their country, and since it is a consumer-driven market, big brands in the developed countries dictate the scenes. Due to the hefty fine, a number of manufacturing units, even in the developing countries, are following suit and analyzing and testing their products before their consignment is shipped to the exporting countries. This has given rise to a number of internationally accredited laboratories which offer the test certification conforming to the standards (OEKOTEX Standard 100 and like) required by various reputed brands. Currently, as far as the United Kingdom is concerned, it imposes a 5000 pound fine and 6 months prison sentence for violation of the EU ban on azo dyes (http://2014.igem.org/wiki/images/2/29/Goodbye_Azo_Dye_POSTnote.pdf).

8.4 Ecofriendly textile manufacturing and processing

It has become necessary to take into account protection of the surrounding environment while manufacturing a product which is ecologically friendly and safe to use. The textile industry makes use of various chemicals in pretreatment, coloration, and finishing, and consumes a large quantity of water and energy. In addition, a lot of chemical fertilizers and pesticides are used during the production of natural fibers, which serve as raw materials for the textile industry. Many of these processes give rise to a lot of effluents consisting of unused auxiliaries, colorants, finishing chemicals, heavy metals, etc. It is required that these effluents are made safe for the environment before their delivery to rivers or seas.

In order to make textile manufacturing and processing ecofriendly, a lot of newer and safer alternatives are substituted for objectionable colorants, chemicals, and heavy metals. This has been possible due to the research and development efforts of various institutions throughout the world making ecofriendly textile manufacturing and processing a reality. Wherever the safe substitutes were not available, a safer limit of use of still-objectionable colorants and chemicals is predetermined and legislated accordingly. Beyond this tolerance limit, however, neither the product nor the process can be called ecofriendly.

It is here we need an independent body certifying the product and processes, so that not only the brands which sell their products are assured of such quality and authenticity, but the consumers are also assured and comfortable in accepting the ecolabels on their products. The ecolabels certify azo dyes based on prohibited amines, pesticides, formaldehyde content, heavy metals, chlorinated phenol, benzenes and toluene compounds, phthalates, and organo tin compounds. Depending upon the specific ecolabels, there are specific tolerance limits fixed by the independent bodies within which the products and processes should comply. Some of the ecolabels include OEKOTEX Standard 100 from Germany, Ecomark from Japan, EU flower, Ecomark India, Global Organic Textile Standards (GOTS), etc.

Ecolabeling is based on two criteria:

1. Product based, which consists of Group I–Group IV products as mentioned in OEKOTEX Standard 100.
2. Process based, which suggests compliance to general recommendations.

8.4.1 OEKOTEX Standard 100

The main purpose of standards is to make the manufacturer as well as the consumer aware of the adverse effects, if any, during the fabrication and use of the product on the environment (plants, animals, and aquatic life) as well as on human life. Hence, considering the various products available for speciality performance, the possible dangers have to be highlighted. The testing methods suggested here will enable the manufacturers and brands as well as consumer-rights organizations to ascertain that these products conform to environmental and health standards. Although the manufacturers always claim that their products cause no hazards to human life or the environment, it is important that the brands which sell them and the consumers are doubly sure that indeed such claims are authentic and verifiable. Hence, independent certification or accreditation is necessary to create this trust among these various stakeholders. Presently, when the trade is globalized where demand is in one part of the world, procurement of the raw materials for such products is made in another part of the world, and actual manufacturing and processing is done in another part of the world, it is imperative that such independent mechanisms of certification of quality and compliance which are internationally accepted do exist, and among many such independent certification agencies, OEKOTEX Standard 100 is one of the most widely accepted labels. It is an independent testing and certifying body which not only examines the raw materials and intermediates, but also the products at different stages of manufacture. As far as high performance textiles are concerned, irrespective of their performance category, in each of these cases it is important that the yarns and fabrics as well as garments are tested for dyes, colorants, auxiliaries, and finishes. There are a number of guiding principles that must be followed while such testing is carried out; for example, to identify if there is any harmful substance used which is illegal or to quantify the proportion of some substances which are legally permitted of use up to a certain limit. It could also be possible to identify some harmful substances still not regulated legally, and some other parameters such as healthcare, etc. The OEKOTEX Standard 100 thus establishes such stringent requirements, which go far beyond the requirements of many national legislations (https://www.oeko-tex.com/en/manufacturers/concept/oeko_tex_standard_100/oeko_tex_standard_100.xhtml).

OEKOTEX testing is based on the impact of the product on human health, based on being in contact with the skin. The products are classified, in general, into four categories, Product Class I constitutes clothing and textile items for babies up to 3 years old consisting of bed linen, terry cloth, toys, etc. The Product Class II includes textiles/clothing used next to the skin such as underwear, t-shirts, and bed linens, whereas Product Class III is for outer clothing such as jackets, coats, etc. Product Class IV, however, is for furnishing material consisting of table cloths, curtains, upholstery, etc. The OEKOTEX Standard 100 requires compliance of not only the major raw materials used in the certified product, but also other components such as sewing threads, linings, prints, as well as non-textile accessories such as buttons, rivets, zips, fasteners, etc.

8.4.2 Sustainable textile production

Besides the product, today the business is seen in terms of its sustainability at various production chains and it should encompass not only environmental protection, but also social accountability—accountability to the work force as well as to the consumers. A new OEKOTEX Certification System for various production sites can be used by the brands, retailers, and suppliers to convey to consumers that their production is completely sustainable. It guarantees environmentally friendly production processes, and in relation to social accountability, it assures optimum health and safety conditions and a socially acceptable work atmosphere. It applies to fiber production, spinning and weaving, and finishing facilities, as well as garment manufacturing. As of Jun. 2013, OEKOTEX Certification for production sites was replaced by the Sustainable Textile Production (STeP) Certification label, and its responsibility is with the OEKOTEX Association, with 16 member institutes present in over 60 countries (https://www.oeko-tex.com/en/manufacturers/concept/sustainable_textile_production_step/step.xhtml).

8.4.3 Process based ecolabeling

In this case, the emphasis is placed more on making the process of manufacturing more ecofriendly, and it suggests compliance with general recommendations. It includes recommendations in process change such as using no hypochlorite-based bleaching or chlorinated carriers in polyester dyeing, avoiding the use of dyes based on banned amines and chlorinated benzene and toluene, use of only alkylphenol ethylene oxide (APEO) and nonylphenol ethylene oxide (NPEO) free wetting and scouring agents, and avoiding ethoxylate-based and phosphate containing surfactants, avoiding use of pentachlorophenol (PCP) and copper sulfate as preservatives for size material, and as much as possible, carrying out pretreatment with enzyme desizing, bioscouring, and hydrogen peroxide bleaching. On the dyeing side, use of high exhaustion dyes and those free of banned amines is recommended. Phosphate-based chelating agents are to be avoided, and the method of Right First Time dyeing is recommended; coloration with natural dyes without the use of heavy metal as mordant is also one more recommendation. In finishing, all formaldehyde-based finishing agents are to be avoided, and the fumes of ammonia and benzaldehyde, and acetic acid should be replaced by formic acid. The flame retardants based on bromo ether and chlorofluoro carbons are to be avoided. In printing, water-based inks should not contain PVC (polyvinyl chloride) or phthalates.

In order to protect the environment and have products with the smallest carbon footprints, there are also recommendations for production of natural fibers and synthetic fibers. For example, natural fibers like organic cotton, organic silk, and organic wool, and manufactured fibers such as bamboo and rayons like Tencel and those obtained from recycled plastic bottles are recommended. Although cotton is the most versatile and skin friendly fiber, because it necessitates the use of large amounts of water and huge inputs of fertilizers and pesticides, its carbon footprint is very large, and hence organic cotton grown without the use of toxic pesticides and fertilizers is increasingly advocated. GOTS certify such material and also recommend subsequent ecofriendly pretreatment, coloration, and finishing processes so that the final product can qualify for GOTS certification, which is discussed in the later part of this chapter.

Organic silk is obtained from silk worms which are not killed, and feed on the leaves of the plant which makes use of organic fertilizers. Organic wool is obtained from goats and sheep raised according to accredited organic standards, and is obtained from sheep into which no hormones are injected, and that graze on land where no pesticides are used. Among manmade fibers, the regenerated fibers such as Lyocell/Tencel are considered ecofriendly since the pulp used in making such fibers is sourced from Forest Stewardship Council (FSC) certified forests. Polyester is petroleum-based and not biodegradable, and contains large amounts of antimony used during its polymerization. However, the recycling of polyester drastically reduces the energy and dependence on oil, as well as enables waste utilization and thus indirectly contributes to environmental protection. In addition, chitosan fiber obtained from shrimp and crab shells, as well as recycled polyester obtained from plastic bottles, offer positive contributions to environmental protection.

8.4.3.1 Hazardous chemicals

While discussing ecofriendly textile processing, it is important to know the ill effects and environmental hazards caused by certain conventional chemicals and auxiliaries used in processing of textiles so that their substitution or regulation with respect to safe limits of quantity can be well understood.

Nonyl phenols and octyl phenols such as nonyphenolethoxylates used as surfactants in washing and coloration, being toxic to aquatic life have the tendency of persisting in the environment and of accumulation in body tissues, posing danger to aquatic life. They have been banned since 2005 (Mahaptra, 2015) in the European Union. Phthalates used as plasticizer in PVC and some of the latexes and binders are toxic, mainly DEHP (di 2-ethylhexyl phthalate) as well as DBP (dibutyl phthalate) and BBP (benzyl butyl phthalate) are toxic to reproduction and are under the potential banned list of EU REACH legislation effective from 2015. Flame retardants based on bromine and chlorine are persistent and bioaccumulating, and poly-brominated diphenyl ethers (PBDEs) are capable of interfering in growth of sexual hormones. Under the EU water law, they are listed as a priority hazardous substance. The EU has already banned azo dyes having the potential to release amines which are carcinogenic. These dyes are forbidden to use on textile materials since their contact with skin may result in the reduction of dye chromophore due to the perspiration releasing such carcinogenic amines which are likely to enter the human body through the skin. The anti-odor compounds having antibacterial and antifungal property are normally used in the next-to-skin garments such as socks, sportswear, etc. They are based on tributiltin (TBT), which has a tendency to persist in the environment and build up in the body, and can adversely affect the reproductive and immune systems of human beings, and under EU regulation of the pollution of surface water, products containing more than 0.1% of such organo tin compounds are banned across the EU. PCP and its derivatives, being biocidal in nature, are highly toxic to human and aquatic organisms, and banned in the EU. Trichloroethane used in cleaning the fabric causes ozone depletion and affects the central nervous system, liver, and kidneys. Chloro benzenes used as carriers and solvents in textiles have a bioaccumulative effect, and they affect the central nervous system, liver, and thyroid. Hexachlorobenzene (HCB) and pentachloro benzene (PCB) are "Priority Hazardous Substances" and should be eliminated from surface water. Perfluorinated chemicals (PFCs) used as water and stain repellents in the textile context are difficult to degrade and persist for a long period in the environment. Among them, perfluorooctane sulfonate (PFOS) is a persistent organic pollutant, and is banned in Europe and Canada. They adversely affect the liver and act as a hormone disruptor.

8.4.3.2 Heavy metals

Although some metals and minerals are useful for the health and growth of animals and human beings, many of the heavy metals, due to their toxic nature, pose a great danger to varied aquatic life, as well as human and animal life. The proportion of these heavy metals, if exceeded in the human body, causes disastrous effects on human health, which has already been reported in the literature. While on one hand the deficiency of some of these metals does pose problems, sometimes their adverse effects due to the toxicity of the metals overweigh the effects in favor of limiting these metals in the human body. There are a number of analytical methods of quantitatively tracing these heavy metals in different textile products which are likely to find ways into the human body due to such products coming in contact with the skin. These metals are in general analyzed using atomic absorption and emission spectrophotometer (AAS and AES) and this method is applicable for the metals like Li, Be, B, Na, Mg, Al, P, K, Ca, V, Cr, Mn, Fe, Co, Ni, Cu, Zn, As, Se, Si, Mo, Ag, Cd, Sn, Sb, Ba, Hg, Tl, Pd, Th, etc. Thirty naturally occurring elements are known to be toxic, and the textile product or the process effluent carrying these elements in trace quantities beyond the permissible limit poses danger to the environment and also human life. Quantitative estimation of these elements on the textile products or process effluent is done by using AAS based on flame or graphite furnaces (Ortega, 2002; Saracoglu et al., 2003; Lopez-Molinero et al., 2007; Ghaedi et al., 2007; Sahin et al., 2008). More details with regard to the principles of this technique and its application are described in the literature (Ebdon et al., 1998; Tsalev, 1998; Dulski, 1999).

Heavy metals such as cadmium, lead, mercury, and chromium used in certain dyes and pigments cause damage to the nervous system and kidneys. They have an accumulating tendency in the body. Lead and mercury damage the nervous system, whereas cadmium damages the kidneys or causes cancer. Chromium VI is toxic even at a low concentration to aquatic organisms, and thus the first three metals out of the four fall under the EU's "Priority Hazardous Substance" list.

The studies of quantifying the toxic metals in cotton- or polyester-based products have been reported in the literature (Saracoglu et al., 2003; Grabaric et al., 1999; Tüzen et al., 2004). The dissolution of the sample containing these metals is an important step, and recently a microwave-based technique that is fast and reproducible is gaining increasing importance for the preparation of the samples for such studies (Narin et al., 2004; Tuzen et al., 2008). The concentration of textile samples dyed with different reactive dyes containing a varied extent of trace metals obtained from Turkish mills were determined using similar AAS studies and were compared with the threshold limit values given by the OEKOTEX Standard 100. Although the other elements were found to be within the limit, the copper and cadmium contents were found to be higher in the pigment-dyed material (Onetti and Innocenti, 2009). Since the heavy metals are likely to be extracted due to the perspiration of the human skin, similar studies are also carried out to determine the efficacy of extraction in the artificial perspiration (http://news.bbc.co.uk/2/hi/europe/4524772.stm).

8.5 Restricted substance list

Restricted substance list (RSL) is prepared as per the EU ecolabel for textile products. The OEKOTEX Standard 100 certification ensures compliance to the list and the threshold limit, as well as verification and testing methods. Under the substances of very high concern (SVHCs), textile products like *elastins and acrylic* used for babies and children under 3 years of age have a limiting *N,N*-dimethyl acetamide value for 0.001% *w/w*, and in all other products, the threshold value is 0.005%. The actual testing can be carried out by subjecting these products to solvent extraction and analyzing the extract using GC-MS or LC-MS (https://www.wewear.org/industry-resources/restricted-substances-list/; http://levistrauss.com/wp-content/uploads/2014/09/July-2014-RSL-English.pdf).

In the case of *biocides* giving antibacterial property to the textile material based on nano silver, nano zinc, triclosan, tin organic compounds, benzimidazole derivatives, etc. the limit of detection is 1 ppm for individual substances. The testing method for them is quite complex, requiring derivatization of the same with acetic anhydride, and analysis on capillary GLC with electron capture detection. These antibacterial agents and biocides are also used in natural fibers during their transportation and storage. In chloro phenols, organo tin compounds such as TBT, TBDOT, polychlorinated biphenyls, etc., their cumulative detection limit is 0.05 ppm, and in the case of dimethyl fumarate, it is 0.1 ppm, and most of the time the supplier of these products has to declare the nonuse of such chemicals. In textile chemical processing, a number of auxiliaries and surfactants are used as explained earlier, and the permissible limit values of alkyl phenols present in them is up to 50 ppm. The test method is C65 Solvent extraction HPLC-MS.

Halogenated carriers which are used to accelerate dyeing of polyester, acrylic, and polyamide such as 1,2-dichlorobenzene, 1,2,4-trichlorobenzene, etc. should not be used since the limiting value is 1 ppm and material safety data sheet (MSDS) is required to be provided by the supplier of dye carriers for verification. Testing of such chemicals is done using DIN54232-2007 or Solvent extraction and GC-MS. The azo dyes with the potential to release the aromatic amines which are carcinogenic in nature are banned for use on textile fibers. The limiting value for each of these amines is 30 ppm and of such banned dyes is 50 ppm. The test method to determine these amines quantitatively is EN14362-1 and 2 and in general the dye suppliers have to provide MSDS to the textile manufacturers.

Chrome mordant dyes are normally used in dyeing of wool and polyamide fibers. The limit prescribed for them is 3 ppm and the testing method suggested is EN ISO 17075:2007. The dyestuff manufacturer should supply MSDS for verification.

Metal complex dye based on Cu, Cr, and Ni for dyeing of wool, polyamide, and blends with manmade cellulosic fiber should only be permitted in dyeing, and provided MSDS of such dyes is supplied for verification.

As far as *heavy metals* are concerned in the Product Class I, that is, for babies and children under 3 years old, the permissible limit of extracted heavy metals in the metal complex dyed textiles are as shown in Table 8.1.

Table 8.1 Permissible limit of extracted heavy metals in the metal complex dyed textiles in Product Class I

Sb—30 ppm Pb—0.2 ppm	As—0.2 ppm Cr—1 ppm	Cd—0.01 ppm Ni—1 ppm	Co—1.0 ppm Hg—0.02 ppm	Cu—25 ppm

The limit values for all other product classes, including interior textiles recommended, are shown in Table 8.2.

Table 8.2 Permissible limit values for all other product classes including interior textiles

Sb—30 ppm Cu—50 ppm	As—1 ppm Pb—1 ppm	Cd—0.1 ppm Ni—1 ppm	Cr—2 ppm Hg—0.02 ppm	Co—4 ppm

Normally the products are tested by extracting the heavy metals using the extraction method—DIN EN ISO 105-E04-2009 (Acid sweat solution) and the detection is done by GC-ICP-MS.

In case of *printing paste*, more than 5% *w/w* volatile organic compounds (VOCs) should not be permitted. These include acrylates, styrene monomers like acrylamide, butadiene, polyols, acrylonitrile, formaldehyde, hydrocarbons, aliphatic hydrocarbons (C10–C20), ammonia, etc. MSDS for printing formulation has to be supplied, and the testing can be done by solvent extraction and GC-MS methods. There is a restriction on the use of plastisols in printing binders and restricted phthalates including PVC. The threshold limit is 0.1%.

For the *easy care finishes* the residual formaldehyde should not be more than 16 ppm for babies and 75 ppm for adults in the next-to-skin garments; however, for all other products, including interior textiles, the limit is 300 ppm. The test method followed is EN ISO 14184-1.

Water and oil repellent finishes, normally perfluoroalkyl sulfonate (PFAS) and perfluorooctane sulfonate (PFOS), are not to be used, and their threshold limit is 1 μg/m^2 each in the coated material whereas all other PFAS forms, the threshold limit is 20 μg/kg and MSDS is required to be supplied for verification. The test method is based on solvent extraction and GC-MS or HPLC-MS. For per fluoro carboxylic acids and salts (PFCA), the limiting value is 0.05 ppm or sum total is 0.1 ppm.

The *flame retardants* based on bromophenyl ether or phosphinoxide or chloro ethyl phosphate should not be used, and the limiting value for their presence is 5 ppm. Their testing is done by solvent extraction and GC-MS/LC-MS.

In the case of *coated and laminated textiles*, use of polymers containing phthalate such as DBP, BBP, etc. should be avoided. The threshold limit for such compounds is 0.1% and testing is done by the DIN EN 15777:2009-12 method. PFCAs used here should not be more than 0.5% *w/w*.

The *metal accessories* used in garments should not contain migrating Ni in the metal alloys more than 0.5 μg/cm^2/week when in contact with the skin. The chrome content in the chromium coated accessories has the threshold limit of 60 ppm chromium. For Ni migration, the test method suggested is EN 12472-2005 and EN 1811-1998+ A1-2008. The general phthalates should not be in the plastic accessories, and their limiting value for some total is 0.1%. It can be tested by the DIN EN 15777:2009-12 test method.

8.6 REACH

The European Union has come out with the regulation on chemicals and their safe use which is known as the Registration Evaluation Authorization and Restriction of Chemical Substances (REACH). The registration has to be done with the newly formed European Chemical Agency (ECHA) based in Helsinki, Finland. REACH gathers and regulates the chemicals which may be carcinogenic, mutagenic, or potentially hazardous and accordingly identifies the risk management measures for safer use. This registration is mandatory if the manufacturers are marketing and supplying the chemical to the European market, and the depth of information on such chemicals is linked to the volume of the concerned chemical. In an 11-year period of implementation of REACH legislation, companies preregistered almost 1,43,000 chemical substances marketed in the European Union by 1st Dec. 2008 qualifying for the 10-year period to complete the full registration process. Those supplies of non-REACH chemicals/substances are considered to be illegal in EU countries. Where the hazardous nature and volume are higher, the deadline for registration was Dec. 2010. For moderate-volume substances it was Jun. 2013, and for low-volume, the deadline is Jun. 2018. Those substances manufactured or imported in quantities more than 1 ton per year in EU countries should be accompanied with the required information for registration. It has to indicate the technical information and also risk management measures. When such volume is more than 10 tons per year, an additional chemical safety report is also required to be submitted. The evaluation allows the regularity authorities to decide on whether to further test and assess the information provided by the industry if there is any risk to health or the environment. The SVHC carcinogens and those toxic to reproductive systems and that are bioaccumulative and toxic require authorization. The "CH" of the REACH applies to chemicals consisting of raw materials, intermediates used in day to day life, clothes, etc. The requirement of REACH is communication of information on chemicals forward and backward in the supply chain in relation to the health and the safety effects of the products used. REACH came into force in Jan. 2009, although its first reading was done in European parliament in Nov. 2005 (Onetti and Innocenti, 2009). The complete enforcement is expected by 2015. Since it is based on a commitment to protection of human health and the environment, the toxic substances such as phthalates and brominated flame retardants are being phased out under REACH. In order to enforce tight control, the ECHA publishes the REACH authorization list about SVHC. Although at a very low level some of these chemicals are not health hazards, many of them being bioaccumulative in nature exceed that limit, and may also react with another substance, creating a newer risk.

8.7 Organic textile standards

Presently among the various accreditation agencies, the GOTS is considered to be the widely accepted global standard for textile products using organic fibers. It not only seeks a high level of environmental compliance along the entire supply chain of organic textiles, but it also requires a commitment to social accountability. The fibers need to be organically harvested with social responsibility and environmentally friendly processes. Additionally, at every stage of the supply chain, necessary care is required to be taken until the final GOTS label is put on the product, which provides consumers, globally, with credible assurance. The international working group of GOTS comprises member organizations such as the Organic Trade Association (OTA, USA), the International Association of Natural Textile Industry (IVN, Germany), the Soil Association UK, and Japan Cotton Organic Association. GOTS certification is offered for organic farming and environmentally friendly and socially responsible textile processing either by inspection on sites or through independent, specially accredited bodies. GOTS certification confers the hallmark of quality to environmental textiles. Such certification is offered to textile processing, and manufacturing, as well as to trading agencies. Certification and assessment of all inputs and accessories used is carried out. An environmental management system consisting of waste water treatment is verified. Inspection of social criteria management and risk assessment is done and the organic nature of the fiber is traced. The key norms of the International Labour

Organization (ILO) have to be followed to conform to social criteria first, then the textile effluent is discharged to the surface water, and its chemical nature before and after treatment has to be ascertained, and finally, the process must conform to environmental norms. The manufacturers as well as processors having such GOTS certification will surely find a competitive edge in getting orders from the brands which are convinced that consumers are ready to pay a premium for such products. GOTS is certification for operations and products. Anyone buying a textile product needs to emphasize such certification from the suppliers. There are two GOTS label grades that exist—"Organic" or "made with organic". When GOTS certifies as "organic" it requires a minimum 95% of certified organic fibers. For GOTS certification "made with organic", a minimum of 70% of the fiber must be organic in the product, and in such a product, a maximum of 10% of synthetic fiber can be used; whereas in the case of socks, leggings, and sportswear, up to 25% synthetic fiber is allowed. For environmentally friendly textile wet processing and manufacturing with GOTS certification, the input chemicals, dyes, and auxiliaries should meet the permitted limit of toxicity and biodegradability, while the toxic heavy metals, formaldehyde, GMO, and aromatic solvents are banned. The packaging material should be PVC free. GOTS latest version 4.0 tallies with very active Greenpeace Detox campaign's 11 input groups, which are prohibited. They are shown in Fig. 8.2.

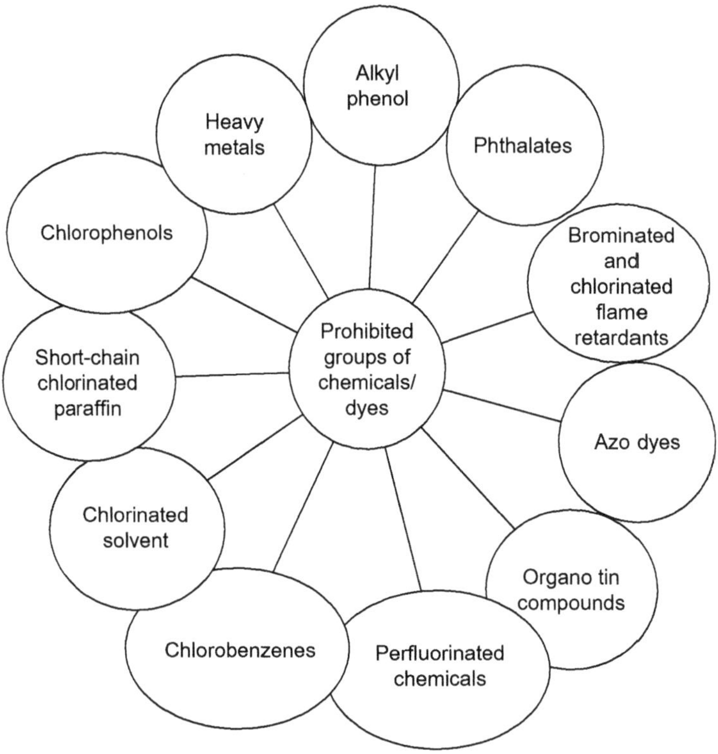

Fig. 8.2 Prohibited groups of chemicals under GOTS.

The quality assurance through GOTS certification is based on the assessment of MSDS. Each supplier has to supply such GOTS certification for their input raw materials and chemicals along with MSDS so that the final product can also qualify for GOTS certification. GOTS limit values and the test methods are quite relevant to Greenpeace Detox chemicals (http://www.global-standard.org/certification.html). The public database of GOTS enables users, as per their needs, to search GOTS certified products, entities in the supply chain, etc.

A number of facilities certified to GOTS increased to 3663, which is an 18% increase over the last 2-year period, and they are presently spread over 64 countries. India has the largest share of this increase, adding 338 facilities in year 2014, and Bangladesh 89, Germany 32, and Turkey 21. The first International GOTS conference in Mumbai in 2015 was attended by more than 250 delegates from 12 different countries. This indicates an increasing awareness among the consumers as well as various stakeholders of textile and garment manufacturers with regard to environmentally friendly processing with social responsibility.

8.8 Concluding remarks

Sustainability issues comprised of ecofriendly processing/manufacturing of products, as well as social accountability to the workforce and the consumers, are becoming increasingly important in the consumer-driven market, the brands, retailers, and manufacturers are giving serious consideration to the accreditation and ecolabeling of their products and processes. While these brands would get a premium for their products, in general, the workforce in the textile value chain would also get a better working atmosphere and proper wages, and the environment will get security and protection from manufacturing activities. Such a win-win situation will indeed contribute to better value for the money for consumers at large and to the true prosperity of society.

References

Chequer, F.M.D., Dorta, D.J., Palma de Oliveira, D., 2011. Azo dyes and their metabolites: does the discharge of the azo dye into water bodies represent human and ecological risks? In: Hauser, P. (Ed.), Advances in Treating Textile Effluent. InTech, ISBN: 978-953-307-704-8.

Chequer, F.M.D., Oliveira, G.A.R., Ferraz, E.R.A., Cardoso, J.C., Zanoni, M.V.B., Oliveira, D.P., 2013. Textile Dyes: Dyeing Process and Environmental Impact. InTech.

Dulski, T., 1999. Trace Elemental Analysis of Metals. Marcel Dekker, New York.

DyStar Customer Service, 2000. Ecology. News Letter (2). DyStar Textilfarben GmbH & Co. Deutschland KG, Customer Service Ecology.

Ebdon, L., Evans, E.H., Fisher, A., Hill, S.J., 1998. Analytical Atomic Spectrometry. John Wiley & Sons, Chichester.

Ghaedi, M., Ahmadi, F., Soylak, M., 2007. Preconcentration and separation of nickel, copper and cobalt using solid phase extraction and their determination in some real samples. J. Hazard. Mater. 147, 226–231.

Grabaric, Z., Bokic, L., Stefanovic, B., 1999. Determination of iron in raw materials, during fabric processing, and in wastewaters of the textile industry. J. AOAC Int. 82, 683.

Hunger, K., Sewekow, U., 2003. Health and safety aspects. In: Hunger, K. (Ed.), Industrial Dyes; Chemistry, Property, Applications. Wiley-VCH, Weinheim, pp. 625–641.

LGC, 1999. The risk of cancer caused by textiles and leather goods coloured with azo dyes. Report. Presented in CSTEE Plenary Meeting, Brussels.

Lima, A.D., Bazo, R.O., Salvadori, A.P., Rech, D.M., de Palma Oliveira, C.M., de Aragão Umbuzeiro, G., 2007. Mutagenic and carcinogenic potential of a textile azo dye processing plant effluent that impacts a drinking water source. Mutat. Res. 626 (1–2), 53–60.

Lopez-Molinero, A., Calatayud, P., Sipiera, D., Falcon, R., Liñan, D., Castillo, J.R., 2007. Determination of antimony in poly (ethylene terephthalate) by volatile bromide generation flame atomic absorption spectrometry. Microchim. Acta 158 (3–4), 247–253.

Mahaptra, N.N., 2015. Textiles and Environment. Wood Head Publishing India Pvt. Ltd, New Delhi. pp. 42–49.

Narin, I., Tüzen, M., Soylak, M., 2004. Comparison of sample preparation procedures for the determination of trace heavy metals in house dust, tobacco and tea samples by atomic absorption spectrometry. Anal. Chem. 94, 867.

Onetti, C.T., Innocenti, R., 2009. Determination of heavy metals in textile materials by atomic absorption spectrometry: verification of the test method. AUTEX Res. J. 9 (2).

Ortega, R., 2002. Analytical methods for heavy metals in the environment: quantitative determination, speciation, and microscopic analysis. In: Sarkar, B. (Ed.), Heavy Metals in the Environment. Marcel Dekker, Inc, New York, pp. 35–68.

Puntener, A., Page, C., 2004. European ban on certain azo dyes. Quality and Environment. TFL.

Puvaneswari, N., Muthukrishnan, J., Gunasekaran, P., 2006. Toxicity assessment and microbial degradation of azo dyes. Indian J. Exp. Biol. 44 (8), 618.

Sahin, U., Kartal, S., Uelgen, A., 2008. Determination of heavy metals at sub-ppm levels in seawater and dialysis solutions by FAAS after tetrakis(pyridine)-nickel(II)bis(thiocyanate) coprecipitation. Anal. Sci. 24 (6), 751–757.

Saracoglu, S., Divrikli, U., Soylak, M., Elci, L., Dogan, M., 2003. Determination of trace elements of some textiles by atomic absorption spectrometry. J. Trace Microprobe Tech. 21, 389.

Tsalev, D.L., 1998. In: Meyers, R.A. (Ed.), Encyclopaedia of Environmental Analysis and Remediation. John Wiley & Sons, New York, pp. 1583–1605.

Tüzen, M., Sari, H., Soylak, M., 2004. Microwave and wet digestion procedures for atomic absorption spectrometric determination of trace metals contents of sediment samples. Anal. Lett. 37, 1925–1936.

Tuzen, M., Onal, A., Soylak, M., 2008. Determination of trace heavy metals in some textile products produced in Turkey. Bull. Chem. Soc. Ethiop. 22 (3), 379–384.

Design, evaluation, and applications of electronic textiles*

9

H.L. Wainwright
E-Textile Technology Consultant to fashion designers, Monessen, PA, USA

It would be appropriate to begin with a brief summary of the evolution of modern functional textiles into responsive E-Textiles:

- 1985 LED/Optic displays (H. Lee Wainwright)
- 1998 Conductive thread (Amberstrand® fiber)
- 1997 LED/Optic CNC production machinery (Wainwright/ANI-Motion, Inc.)
- 2005 Interactive hugging shirts (Cutecircuit®)
- 2005 Photovoltaic (PV) energy-producing fibers (Konarka Technologies)
- 2006 Lumalive® programmable LED array shirts (Philips)
- 2006 Bio-sensor feedback fabrics (Wainwright/Exmovere®)
- 2006 IR IFF LED/Optic E-Textile & sleeve digital display (H. Lee Wainwright)
- 2010 IR energy output fabrics (Celliant®)
- 2013 Programmable OLEDs embedded into textiles (Cutecircuit®)
- 2015 Programmable E-Textile wall panels and rugs (Philips)
- 2015 WIFI/Bluetooth color-controlled LED/Optic E-Textiles (H. Lee Wainwright).

9.1 Fundamental design issues for incorporating electronics into fabric environments

Unlike smart fabrics that contain chemicals or compounds that behave in predetermined ways in response to external stimuli such as light, heat, and moisture, electronic textiles (or E-Textiles), contain one or more of the following elements:

- LEDs or OLEDs
- Switches
- Electroluminescent (EL) wires
- Wires
- Conductive threads
- Sensors
- Power sources
- Printed circuit boards (PCB)
- Microprocessors
- RF receivers
- IR receivers
- Bluetooth® data links

*Inventor of LED/Optic Fabric Displays, received over a dozen patents in the E-Textile applications and manufacturing sectors, Technology Consultant to fashion designers including the late Alexander McQueen for creating E-Textile fashions. Award Winner: NASA's "Design the Future" contest with E-Textile application, Guest Speaker at Flexible Display, Smart Fabric, Textile, and LED conferences

Performance Testing of Textiles. http://dx.doi.org/10.1016/B978-0-08-100570-5.00010-4

- WIFI data links
- Rigid or flexible PV panels
- Digital cameras
- Digital audio players
- Speakers
- Audio, video, digital, WIFI, Bluetooth®, RF, zigby, or USB inputs/outputs
- Heating or cooling elements.

These elements must be creatively incorporated into apparel in unobtrusive ways while remaining durable enough to withstand normal wear and tear and cleaning procedures. This chapter addresses these challenges with respect to leading-edge technologies that are known at the time of printing. Certainly, however, E-Textile development can be expected to make use of new miniaturization techniques, nanotechnology breakthroughs, user-friendly programmable microprocessors, high-powered flexible rechargeable battery sources, and rugged protective flexible coatings at a rapidly increasing rate.

There are primarily two categories of E-Textiles:

1. The first and simplest applications are embedded "off-the-shelf" consumer electronic devices, such as digital cameras, battery packs, chargers, speakers, headphones, mobile phones, and light emitting diodes (discrete LEDs and LED strips, flexible solar films, digital audio players, EL wires, etc.). These are installed into apparel using standard wiring.
2. The second and most compelling category described in this chapter includes E-Textiles that:
   - Employ micro-processor-controlled fabric color with changing systems that are remote controlled by sensors or smart phones
   - Monitor and display real-time biophysical conditions in the form of digital alphanumeric data
   - Replicate physical motion or mood remotely on another person's apparel
   - Provide responsive color-changing attributes to fabric, depending on a person's mood or in response to surrounding stimuli (ie, sound, temperature, light, lethal gas environments, etc.)
   - Use fabric displays to identify incoming callers on nearby smart phones.

9.1.1 Advantages, disadvantages, and limitations of E-Textiles

As technology advances are made in the field of apparel add-ons, the E-Textiles sector also must be cognizant of potential failure points that could lead to negative impacts. For example, disposal issues will undoubtedly result in new certification requirements and procedures to insure quality control and minimal environmental impact.

Advantages of E-Textiles are somewhat obvious when considering utility value-added technologies that can be combined with textiles:

- monitoring health and fitness (the largest segment of R&D),
- recharging digital devices fitted with lightweight batteries such as LiPo (Lithium Ion Polymer batteries), by using body heat or body movement,
- providing night-time LED safety indicators,
- IFF (Identifying friend or foe) in policing or combat situations,
- indicating acute surrounding environmental hazardous conditions (ie, UV radiation levels),
- displays that link to smart-phone functions, such as emails, texts, voice mails, calls, GPS,
- alerting hospital and first aid staff to medical emergencies,
- using smart-phone-controlled fabric attributes to display utility functions such as voltage, temperature, and other measurements.

Most of the initial phases of research and development of light-emitting E-Textiles seem to have been applied in the arena of costume development for publicity and stage performances, art and design projects funded by arts and science grants, and innovative haute couture fashion shows. However, these incredible eye-popping displays are presently far too expensive to be considered for entry into the consumer market (ie, tens of thousands of dollars each). They are also not built to withstand rugged wear and tear, and they may require special cleaning procedures. The necessary power sources may not last more than a few hours before having to be recharged or replaced.

Fig. 9.1 Janice Martin Couture creation for PIFA (Philadelphia Festival of the Arts 2011). Photo provided by Janice Martin Couture.

As seen in Fig. 9.1, Janice Martin, couture fashion designer, presented a number of E-Textile creations at the Philadelphia Festival of the Arts opening runway show in 2011, with bonded LEDs (Fig. 9.2).

A major challenge facing E-Textiles is the wiring necessary to carry an electric current to the components that may be quite power-hungry. Components need to be protected against shorts or breakage, and should not become entangled during wear or unnecessarily inconvenience the wearer. Several companies have taken on the challenge to meet these issues. Stainless steel, silver, and copper conductive threads are now available from companies such as Amberstrand®, Silverell®, Shieldex®, Boyle Grey®, and Adafruit®. However, this opens new challenges concerning connectivity to components. A major disadvantage is that connections may be exposed to compounds during cleaning that may cause them to corrode, short, or become disconnected through motion. Magnetic conductive connectors offer one solution; however, at present, they are not strong enough to withstand the harsh stress or quick motion normally expected with physical acrobatic movements. This issue is also shared by any type of

Fig.9.2 Bonded LEDs.
Photograph by H. Lee Wainwright.

switching mechanism that uses mechanical switching, and is considered by most to be the "weakest" link in the designing of E-Textile applications. However, led by Despina Papadopoulos and Studio 5050 Ltd one group has created robust connectors designed specifically for E-Textile applications that use durable snaps and flexible flat wires that appear to be particularly appropriate for use with fabrics (Fig. 9.3).

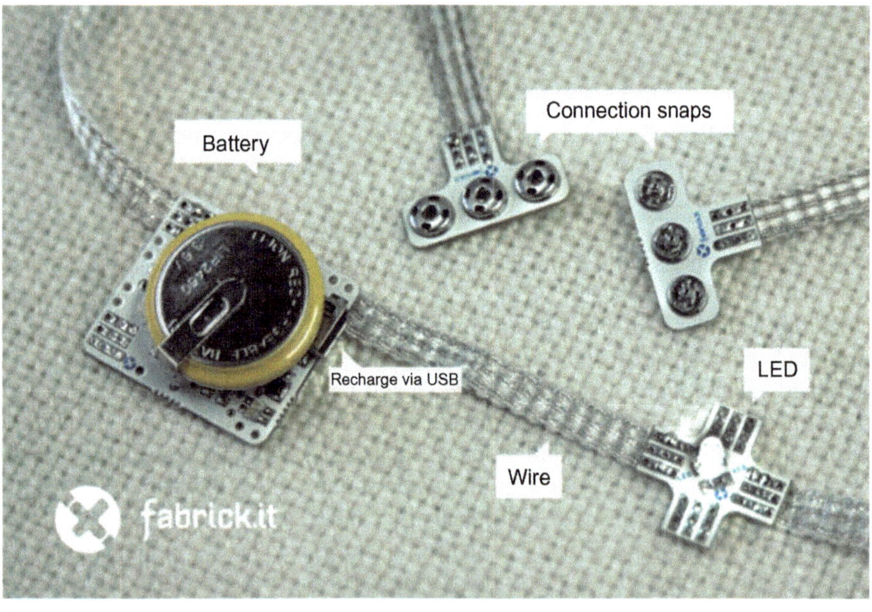

Fig. 9.3 Connectors.
Photo provided by Despina Papadopoulos.

Another research group, System-on-Flex, (TexLab), has been developing creative wiring methods for several years. They have focused on stretchable conductive substrates that allow for movement, eliminating concerns regarding open circuits and broken wiring. Produced with a zig-zag pattern, the wiring expands and contracts with motion. The flexible technologies their team has developed marry well with many E-Textile products made by Sinetra, Texteer, AlarmTextil, Place-IT, LumoLED, TexOLED, Wear-a BAN, TextraLog, Cyber Nomade Suite, Sporty Supaheroe, and more.

However, on the other hand, there are companies that have suffered great financial losses when switches or connections failed in fabric applications as experienced by Fibretronics®, a company that designed mechanical switches permanently embedded inside fabric apparel sleeves and who offered their products to Columbia Sportswear®, a large consumer sportswear company. When the switches were installed in a product that required heating elements, some shorted. This caused overheating and other problematic issues that, in turn, resulted in a multimillion dollar lawsuit (The Seattle Times Mar. 12, 2015, "Judge awards Columbia Sportswear $3.3 million in lawsuit"). This example illustrates the danger of introducing an E-Textile product into the consumer marketplace without extensive testing, and without minimum certification regulations along with independent verification procedures in place. The industry is growing and new safeguards will emerge from the failures that present themselves in the upcoming years. Hopefully, "Burning Man," the yearly event in the Nevada desert in the USA where LEDs and lighting special effects reign supreme, will not become synonymous with E-Textile disasters.

Quality-control testing for each production run of an E-Textile application is absolutely essential for any product that comes in direct contact with an end user.

Companies like the Swiss-based testing facility, SGS, offer testing of E-Textile applications including:

- Physical and mechanical tests
- Chemical (eg, volatile organic compounds—VOC) tests
- Materials tests
- Food contact tests
- Flammability tests
- Performance tests
- Microbiology or Microbiological tests
- Electrical safety tests
- Electromagnetic Compliance tests
- Environmental safety tests.

The most acceptable design criteria for adding electronics to apparel, after safety concerns, should be comfort and style. These two top priorities are needed if wide acceptance in the consumer marketplace is to be obtained. People do not want to tangle themselves up in wires or to have to make adjustments to their normal movements because of additions made to their apparel choices. Some designers have turned to embedding PV flexible solar film into their creations in the hope of eliminating the inconvenience of changing batteries, but that may lack some forethought. Did they consider whether people would want to be exposed to the sun wearing black flexible solar cells for the prolonged periods necessary to charge even the smallest of rechargeable power sources?

Another consideration for designers is the issue of waterproofing an E-Textile application. After all, we live in a wet world, and everything we wear should withstand the weather extremes we normally expect. Waterproofing can be achieved in a number of ways. Adding a waterproof battery case is one approach, but using both a waterproof conformal coating and a waterproof case offers two levels of protection. Conformal coating sprays that are used by the military for protection of electronic devices are readily available from various electronic supply houses. In order to properly certify that an electronic product is water-resistant (safe in rain showers or moisture contact), water-repellent (conformal coated), or waterproof (sealed to withstand underwater environments), an independent lab needs to be involved. Intertek is one such company which offers product testing and certification services and has over 1000 offices located in over 100 countries.

Using hermetically sealed sensors rather than mechanical switches is a good path to follow when designing E-Textile applications. For example, using a vacuum-sealed, magnetic-switch sensor, and the placement of a magnet in a pocket with the sensor located beneath a layer of fabric, offers a switching method that greatly reduces mechanical failure-point concerns. Pressure activated switches are a somewhat better approach than using slide switches that are likely to corrode, rust, wear away, or dissolve after many activations and washings. These "momentary" switches connect using externally applied pressure. Switches do not need to withstand high current loads as a polling circuit that draws micro amps from a coin cell can be used to interrogate the on/off status of a switch to control a series of programmable routines. The same advantage applies to light-sensitive or photon-sensitive activation switches. These methods require a constant drain on a battery, but the current necessary to keep polling these switches to determine if the proper condition is met to cause activation of the display is in micro amps. To put it another way, a small coin cell could power these momentary switches for years without draining enough power to prevent activation.

Most E-Textile light-emitting projects use discrete (single color) LEDs, bonded LEDs (thin insulated clear wire with LED "dots" bonded in line as in Fig. 9.2), or OLEDs (Organic LEDs) that can be added to flexible Mylar substrates to produce curved displays. All of these methods open themselves to motion issues, including breakage and contact of the components and wiring with objects and elements that increase the chances of malfunction due to impacts. E-Textile durability for consumer acceptance is based on function ability being able to survive 50 repeated washings and dryings. The best way to protect these components is to cover them with a translucent protective fabric or material, and diffuse or conduct the light in creative ways to achieve visual attributes beyond the simple light point. One company that has created flexible diffusers for changing the attributes of LED light is Lumitex®. It offers an entire line of flexible diffusers that shape light into diamonds, lines, areas, etc., and that can be overlaid onto LEDs to result in striking designs across surfaces.

EL (EL Lines) can offer a very beautiful and esthetic touch to any costume (Fig. 9.4), but as visually attractive as they appear, there are serious issues when applying this technology to consumer apparel. The systems are quite fragile and not machine washable or waterproof. Any break point causes the entire neon-line effect to shut off. While quite striking for use on dance costumes, good durability is obtained

Fig. 9.4 EL lines on dance performers.
Photograph by H. Lee Wainwright.

only as long as the composite fabric is not bent sharply back upon itself or impacted against a hard object. Many dance companies creatively apply this technology to create exciting displays of "neon-looking" light effects.

There are several companies offering EL panels that can be attached to T-Shirts, but the panels must be removed prior to washing. The panels may be sound responsive or proximity-responsive; that is, when two people wearing paired EL T-Shirts are in close proximity certain areas, the garment may become illuminated or the lighting effect may be decreased as the distance increases. Interactivewear® is one such company offering interactive E-Textile apparel.

It is recommended that EL lines should be purchased from a certified dealer rather than from questionable suppliers on the Internet offering discounted pricing. Although poor quality LEDs from discount houses may not give reduced output, poor quality EL devices may contain black areas and relatively pale lighting effects. EL has another disadvantage compared with LEDs. EL requires 5–20 times the power compared with LEDs (>100+mA compared with 5–20mA for most LEDs). Typically LEDs emit 20–100 times more intense light output and are able to withstand far more punishment without failing.

LED outputs have greatly increased while pricing has inversely dropped over the years. Blue LEDs starting at $50 each in 1985 had an output of <1000mcd (millicandela) requiring 20mA. Today, the same 20mA current is required to light them, but the output intensity has increased by more than 100 times the original brightness to 100,000 mcd. Most dramatically, the cost has dropped to well under a dollar!

As already mentioned, light-emitting E-Textiles presently under development make use of LEDs attached to conductive wires, conductive threads or conductive fabric weaves. The first company to mass produce E-Textiles was ANI-Motion Inc. in 1997. The company is now defunct, due, perhaps, to being a bit too far ahead of the consumer-acceptance curve. The company took a uniquely different approach to developing and producing E-Textile applications, by eliminating concerns over wires breaking or expose of LEDs to the elements. The electronic portions were coated with an insulating layer and buried deep inside apparel, out of the reach of consumers, and

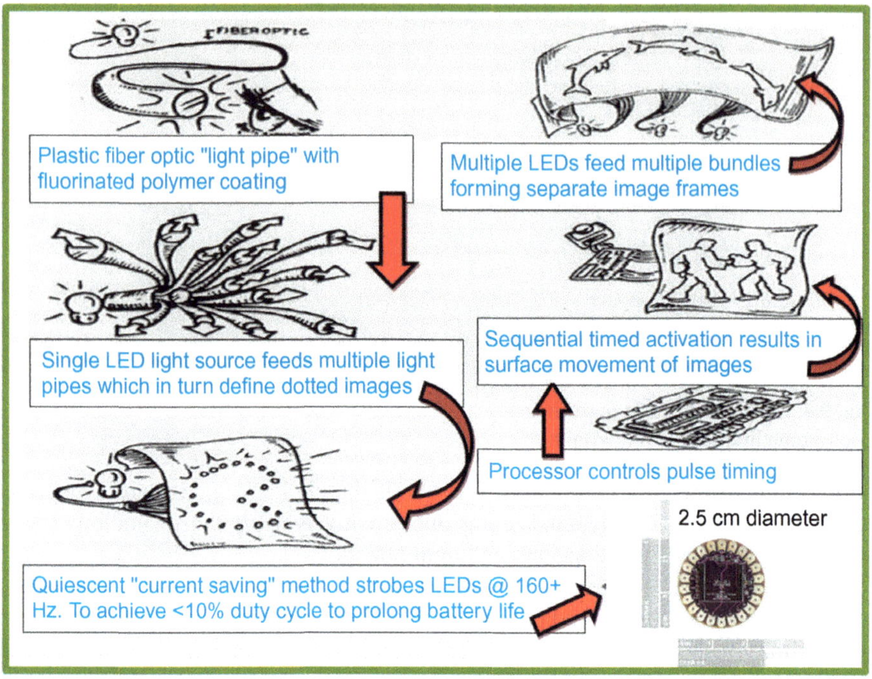

Fig. 9.5 LED display on fabric surface by H. Lee Wainwright.

plastic optical fibers (POFs) channeled the light from LEDs to the surface of the fabric (see Fig. 9.5). The LEDs were controlled using a small microprocessor embedded in the garments.

A detachable leather battery case held a recessed switch to prevent accidental activation. Style and convenience topped the list of design requirements. Ani-Motion manufactured thousands of E-Textile animated jeans, jackets, and costumes for corporate clients, celebrities, theme parks, and retail mass merchants before the year 2000. (Fig. 9.6 shows a bat design created for Jim Steinman, lyricist of the band, "Meatloaf".)

Other LED/Optic displays (Light Emitting Diodes attached to POF) that Ani-Motion manufactured included displaying animated cartoon characters that ran across their backs, and these were sold through Disney parks. The company produced haute couture outfits for the late Alexander McQueen for his GIVENCHY® runway show in Paris, as well as illuminated costumes for Hollywood celebrities. An example was worn by the late Robin Williams (as shown in Fig. 9.7), another by Isaac Hayes at the 2000 Academy Awards, and another by Britney Spears on her World Tour. The company produced the first machine washable pair of jeans with electronic displays—an illuminated rainbow on the rear pocket—that sold through Limited Too®. Later, designer Dalia from Los Angeles sold blue jeans through Macy's with a pair of animated kissing lips under the brand name, American Chi®.

Fig. 9.6 LED/Optic Bat Jacket commissioned by Jim Steinman, lyricist for "Meatloaf." Photograph by H. Lee Wainwright.

Fig. 9.7 LED/Optic shirt worn by Robin Williams at the 2000 Academy Awards Ceremony. Photograph by H. Lee Wainwright.

Only limited numbers of E-Textile apparel products have actually entered into mass consumer markets, but some are fast approaching consumer acceptance. Although LED illuminated sneakers have been around for several years, sporting apparel with LEDs that light for better visual recognition by drivers at night are gaining acceptance. T-Shirts employing EL that respond to ambient sound, have become affordable. However, the T-Shirts still require a user to remove the electronics before machine washing.

9.1.2 Installing sensors, batteries, wiring, and other hardware challenges

Research has been ongoing on making functional inorganic nanocoatings for application onto the surfaces of textiles for the purpose of conducting power to discrete components. Amberstrand® is one of the pioneers that developed a conductive copper thread that could be used in conventional sewing machines for use in fabric structures installed in military helicopters for transmission of low current power to digital devices. However, the fabric containing the wires had to be mounted between insulating layers since the thread wires were unshielded. The fabric could not be sharply folded for fear of shorting, and the copper thread did not perform well under rugged exposure conditions.

Another challenge being undertaken by researchers is the creation of layered organic polymers that are flexible enough to withstand the punishment of being embedded in fabrics to form flexible transistors. These bear many long names such as fiber-embedded electrolyte-gated field effect transistors or organic field effect transistors (OFETs); electric double-layer capacitor-gated organic field effect transistors or EDLC-OFETs; electrochemical transistors; thin film transistors).

Although a fabric transistor may be considered the "Holy Grail" of components for E-Textiles, it will ultimately boil down to manufacturability on a mass scale, and initiating procedures that assure quality, consistency, and durability.

9.1.3 Bio-physical monitoring: Connecting to the cloud

Today, one can slide one's finger over a touch sensitive screen to open up an application on a smart phone or tablet. Companies such as Exmovere Inc. have developed an application that allows one to view the bio-physical condition of a person who is continents away. Wearing specially designed E-Textiles or watches with Galvanomic Skin Response (GSR) sensors that measure the change in resistance or optical sensors that measure the color of the skin enables a microprocessor to output digital data on heart rate, ECG, mood and other biophysical parameters. By employing special algorithms that translate the data, the information can then be sent to a digital display on a laptop or to an E-Textile fabric display that enables viewers to see it even over the Internet. Displays on wearable apparel could be especially useful in alerting workers of impending dangers from gases or radiation in confined spaces such as mines, tunnels, or hazardous environments. Sleeve display monitors could also be used in military situations where real-time biophysical conditions of soldiers may necessitate quick decision making to save lives.

A health monitoring E-Textile technology was demonstrated for the first time in 2006 at the Smart Fabric Conference in Washington, DC, by Exmovere. The medical device company created the software and Bluetooth communication protocols that allowed an E-Textile jacket created by H. Lee Wainwright to display the biophysical status of a person in real time from a watch (Fig. 9.8) installed with GSR sensors.

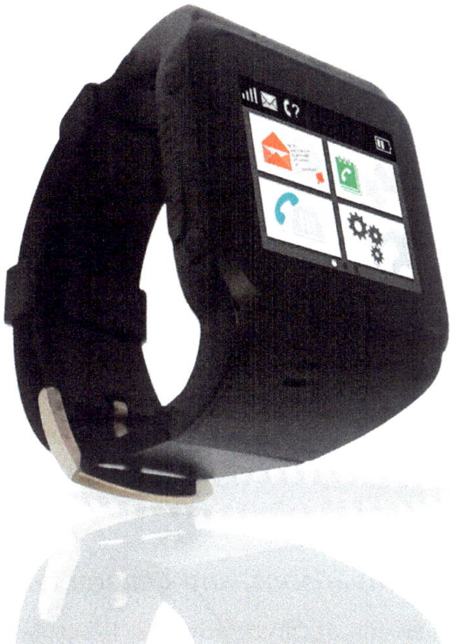

Fig. 9.8 Bluetooth/WIFI enabled mobile phone with bio-physical sensor watch.
Photo provided by Exmovere.

9.1.4 Addressing recycling and disposal issues for E-Textile components

Recycling fabric apparel items today is a process that simply requires taking them to depositories where thrift shops and second-hand retail merchants can resell them. For the most part there are few technical problems associated with recycling apparel. However, this will not be the case with E-Textiles. First, there are requirements already in place that prevent Lithium and other types of batteries from being buried or burned because of their potential environmental impact, so clothing with electronics must be designed in such a way that it is possible to remove and dispose of batteries and/or other elements "potentially harmful" to the environment. European certification clearly states that batteries must be able to be removed from retail products that include apparel. Although at the time of printing, the USA does not require special procedures for disposing of batteries in products, eventually this could change as social pressure demands that more attention be given to environmentally friendly "green product" stamps of approval.

One method to address this issue was implemented by a company that supplied T-Shirts to a large mass merchant more than a dozen years ago. It encapsulated the batteries, circuits, and LEDs in hard, clear epoxy shells that were placed in sealed pockets. These units were capable of flashing up to 100,000 times before shutting down. When these shirts were disposed of, the electronics, batteries, and electrical leads were therefore sealed and were suitable for landfill disposal, but they were not suitable for disposal by incineration. Recycling is a far more environmentally friendly approach and should be standardized as the proper path for disposal. E-Textile goods will need to be labeled indicating exactly how the item is to be disposed of.

9.2 Safety requirements & evaluation criteria for E-Textiles

As E-Textiles evolve, certifications and safety requirements will emerge first from the private sector. However, as the industry matures, government intervention will most likely begin to define detailed requirements for consumer safety, and particularly, for child safety parameters. In one sense, the E-Textile industry, while in its infancy, is more like the Wild West. The "no rules" experimentation, exploitation of new concepts fueled by trial and error, entrepreneurial spirit, creativity, unproven technologies and risk—are now typical of the growing pains of an emerging industry.

9.2.1 Categories, comparisons, and general types of E-Textiles

From the foregoing discussion, it can be seen that E-Textiles utilizing sensors, microprocessors, and technology-enhanced materials can be classified into two basic subcategories:

1. Fashion statements (costumes, catwalks, and consumer entertainment)
2. Utility functions (military, medical, safety, and pragmatic applications, such as charging and controlling digital devices, caller ID and other smart phone applications, and visual feedback transferred to wearable displays).

As already mentioned, category one covers novelty applications that tend to be pricey and presently lack the ability to be manufactured on any massive scale. Over time, some of these creative fashion experiments may appear in consumer markets only if manufacturing methods enable sufficient cost reductions to occur and quality assurance procedures have been implemented. Prototypes such as Lumalive® T-Shirts by Philips Corporation utilizing LED arrays (Fig. 9.9) underneath T-Shirts designed to display alphanumeric and graphic images (Fig. 9.10) face huge challenges when consideration is given to manufacturing them for mass markets. However, Philips® has successfully created a line of unconventional E-Textile LED array surfaces for walls and rugs that enhance environments for decorative convenience and safety which is less cost-sensitive and more utility based.

9.2.1.1 Body heat, flexible PV, motion power generation methods

The following several groups are heavily involved with research revolving around creating a communicative link between a piece of apparel and the person with whom it contacts.

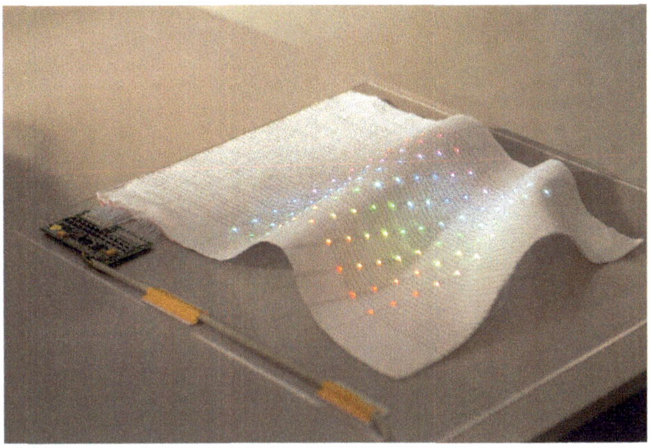

Fig. 9.9 Programmable RGB LED embedded fabric.
Photo provided by Philips Ltd. Peter van den Hurk Editor, writer and content specialist Philips Content Center of Excellence.

Fig. 9.10 Philips "Lumilive" T-Shirts with programmable LED displays.
Photo provided by Philips Ltd. Peter van den Hurk Editor, writer and content specialist Philips Content Center of Excellence.

- Whispers Research Group, a Simon Fraser University SIAT project, is a group incorporating artists, designers, computer scientists, and software engineers involved in developing real-time interactive communication between fabrics and the people wearing them.
- A project called "Karma Chameleon" incorporates a new material technology that changes the fabric attributes according to motion. Deemed "Kinetic Electronic Garments", the fabric integrates a shape memory alloy called Nitinol in substrates resulting in dresses that change shape and color over time using resistive heating and control electronics (Fig. 9.11).

Flexible PVs that produce electricity at low currents (50–100mA) for charging LiPo or NiCad (Nickel Cadmium) batteries are available through a few suppliers by incorporating PV film into an E-Textile application. However, this approach could be considered both

Fig. 9.11 Textile that changes shape in accordance with body heat.
Photograph by Ronald Borshan © 2012 provided by Joanna Berzowska, Design and Research Director XS Labs ©2012.

difficult and somewhat limited because of the inherent incapability of flexible PVs to withstand machine washing due to their inability to withstand flexing, which causes breaks in connections. The other issue is the time necessary to recharge batteries (http://www.powerfilmsolar.com/education/calculations/) (ie, several hours) for solar cells to be in direct sunlight to recharge even a low capacity rechargeable battery, as already mentioned. A better approach might be to install PV film in a banner, awning, or curtain to charge a small battery meant to power a single LED in an LED/Optic application able to light hundreds of points defining the name of a restaurant, product, or power an animated decoration overnight.

9.2.2 MSDS (material safety data sheet) for embedded components

As E-Textiles become more widespread in positioning their appeal to the masses, standards will emerge that outline the necessary safety concerns and quantify them on for retailers to follow. For example, adding a motion switch containing mercury or using soldered contacts with lead alloys rather than silver added to clothing might suggest new regulations should be generated to protect consumers. In order to do this, the MSDS presently used for shipping hazardous materials, chemicals, and other products containing questionable components may present a template from which a new set of rules may be forged to regulate the E-Textile industry.

An MSDS Certification is also required by shipping companies when handling materials with safety concerns. Intertek with offices in 100+ countries is claimed to be able to provide MSDS information.

MSDS covers the following criteria regarding components:

- Carcinogicity (Causes Cancer)
- Corrosion (Dissolves materials including skin)
- Toxicity (Poisonous)
- Skin irritation (Burning sensation)
- Sensitization (Allergic reaction to majority exposed)
- Target organ effects (Signs and symptoms for various bodily organs).

Note that some rechargeable batteries may contain Lithium and Lead-Acid while LEDs contain—Gallium, Arsenic, and Phosphorous. These materials alone could be very hazardous to contact, however, since they are encased or embedded in a product unexposed, an MSDS would not be applicable.

9.3 Summary of present applications

Although the most lucrative and utility-based E-Textile applications seem to be growing rapidly in medical applications, other uniquely fabric-based technologies appear to be burgeoning in other areas, particularly in fashion and entertainment. E-Textiles are beginning to find their place in mass markets around the world.

Designers seeking to create an E-Textile application will need to choose an electronic platform from which to control the functions they desire in their work. The following platforms lend themselves well to beginners and experts alike to begin their projects:

- Adafruit's GEMMA miniature wearable PCB
- Lilypad XBEE radio transceiver for wireless wearables
- Arduino generic PCBs for Bluetooth, Zigby, and RF applications
- Inventables offers preprogrammed LED drivers with standard routines
- Mouser offers the Winbond, PIC chip, and other microprocessors.

9.3.1 Health monitoring and athletic training

Consumers becoming health-conscious are focusing on tracking their own health while coaches are discovering new fabric-based technologies to assist them in athletic training.

1. Consumers

As mentioned earlier, the Exmovere® watch connects users to fabric displays to instantly monitor their vital health statistics, but Reebok® and NIKE® along with over a dozen health monitoring consumer product companies have begun to introduce consumer health monitoring devices. Optical Oximeters along with other sensors in chest bands, wrist bands, and inside sneakers are designed to monitor the skin and transmit data via Bluetooth® to a watch, smartphone, or nearby computer to display and record heart rate along with other biophysical parameters.

2. Working dogs (ie, DEA and bomb detection)

Using GSR sensors in socks attached to a dog's paw or on a collar against the skin on a dog's neck could transmit ECG and heart rate to indicate the physical performance of a dog. By monitoring animal sensitivities and performance data, it should be possible to better train and deploy bomb-sniffing dogs, racing animals, and seeing-eye dogs.

9.3.2 Military applications

"Future Soldier," is a project headed by the United States DOD (Department of Defense) which directs research towards creating a technologically superior soldier with new supportive technologies to add efficiency by increasing the speed of communication and awareness of surrounding battlefield conditions (http://nypost.com/2009/05/23/soldier-of-the-future/). E-Textile research is underway to develop sleeve displays durable enough to withstand battlefield conditions. OLED displays able to curve around surfaces is one avenue of research while another avenue studied by BAE Systems involves using fabric based LED/Optic digital data displays (Fig. 9.12) impervious to short circuits and impact damage.

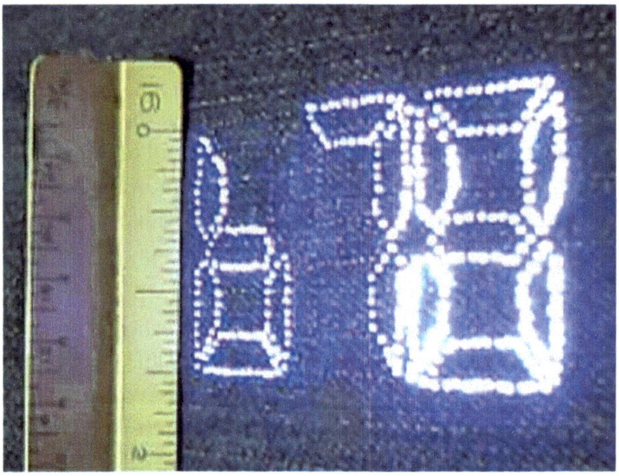

Fig. 9.12 Fabric digital display.
Photo provided by H. Lee Wainwright.

Another technology that may prove worthwhile to investigate would be to mount an LED/Optic display on the nylon covering of a bullet-proof flak vest. Embedding Mitsubishi PMMA (polymethylmethacrylate) POFs across an area on a fabric surface should allow that display to transmit encrypted data via infrared frequencies. By transmitting the biophysical status of someone wearing GSR sensors in a wristband, an observer using an FLIR (Forward Looking Infra-Red) eyepiece, could simply look at an injured soldier to determine his/her immediate state of health. This might prove useful in determining the next immediate response to a life-threatening situation.

In addition, E-Textile communication could be useful in identifying a friend or foe on a battlefield. By employing Infrared LEDs operating at 940 nm, a frequency beyond the eye's ability to see, connected to PMMA plastic fiber bundles able to conduct IR defining alphanumeric characters in a dynamic or static display could be used to instantly determine a person's identity (Fig. 9.13).

Fig. 9.13 IR sensitive camera shows IR LED/Optic output from jacket in photo.
Photo provided by H. Lee Wainwright.

9.3.3 Entertainment and fashion

Although several E-Textile products have entered into the marketplace, none are as visibly eye-popping or memorable as those seen on celebrities, performers, and recording artists. From Wainwright's LED/Optic flaming leather pants worn by Britney Spears on her 2000 World Concert Tour to Cutecircuit's OLED Twitter dress worn by Nicole Scherzinger on the red carpet in 2012, no fashion is as memorable or photographed more than an E-Textile creative fashion statement! Barbara Layne from her studio, subTela, part of the Hexagram Institute at Concordia University in Montreal created the Tornado dress and two interactive jackets with digital displays (Fig. 9.14)

Fig. 9.14 Tornado light-emitting and interactive jackets with digital output.
Photograph provided by Barbara Layne, sbTela, division of Hexagram Institute at Concordia University.

Fig.9.15 Shows an EL reactive dress.
Photograph provided by Amy Konstanze Mercedes Rainbow Winters.

while Amy Winters created an eye-popping E-Textile presentation that responds to smart phone input and ambient environmental stimuli (Fig. 9.15).

9.4 Research paths leading to the future of E-Textiles

E-Textile R&D will continue to create more opportunities for growth industries in the future. Although some developments may sound more like they belong in a science fiction movie, they will fast become an everyday expectation once the manufacturing processes are worked through.

Ongoing research projects:

1. Flexible organic PV's that could be woven into fabric on a loom could provide power to E-Textile applications, charge digital devices, and still remain soft and stylish. This technology has a far way to go since at present the efficiency rating of organic PV cells ranges between 2.79% and 3.27% while solar films and polycrystalline solar panels have attained up to almost 30% conversion of solar to electrical energy.
2. The use of Goretex® "breathable" fabric as a base for screen-printing electrochemical elements that enable a wearer to detect the presence of explosives including TNT. The fabric responds accordingly to alert a wearer of the presence of specific chemicals used in explosive devices.
3. Screen printing could be used to place electrochemical sensors on fabric surfaces. For example, on Neoprene wet suits suitable sensors could detect the presence of environmental pollutants and security threats. Also electrochemical sensors printed onto fabric could directly respond to heart rate, EEG, and ECG bio-physical status. Sensors able to perform chemical analysis could instantly alert a wearer to any danger of exceeding normal healthy limits.
4. Installing LED/Optic displays into apparel and accessories can enable areas on an E-Textile to change colors according to smart phone controls, including digital camera color-matching,

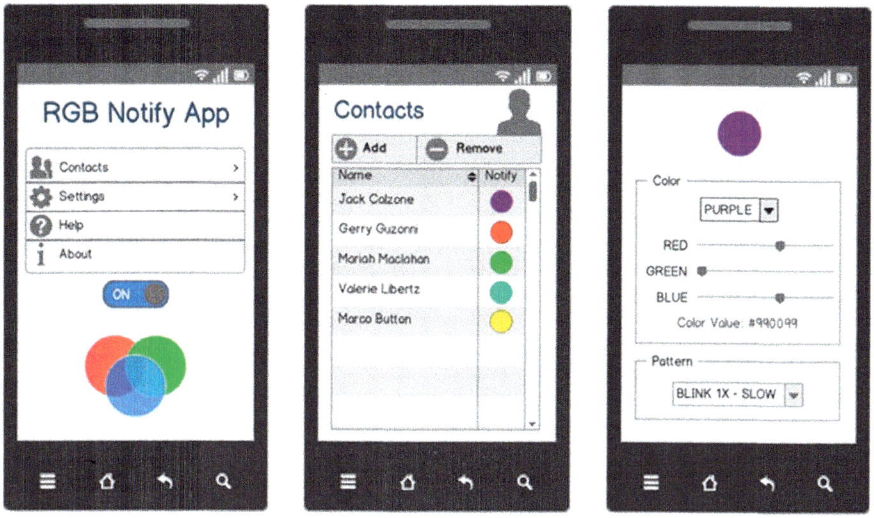

(A) Main Screen (B) Assigning Routines to Contacts (C) Specifying the Contact's Color and Pattern

Fig. 9.16 Smart phone MENU options allow users to adjust the color output of areas on wearable apparel.
Graphic provided by H. Lee Wainwright.

ambient audio responses, caller ID, and wearable programmable routines to match our moods. Fig. 9.16 illustrates a MENU function allowing one to choose how the fabric will respond to a particular phone caller. Fig. 9.17 shows how one may touch a color wheel to change accents or areas on garment surfaces.

5. A group of University of Miami researchers are continuing their development of a flexible wearable surface employing microthin spaced LEDs which display full color images matching surrounding surfaces or are able to be programmed to display custom images. This technology might be particularly interesting to apply to military uniforms for camouflage.

In conclusion, in the future, E-Textile technologies may enable us to adjust our "second skin", so to speak, for comfort, hygiene, protection, and expression like never before imagined.

Finally, fashion statements do not have to be limited to standard designs, but rather it may be possible to create our own apparel coloring instantly at the touch of a smart phone or other digital device. Like a chameleon, we will be able to adjust our appearance to match our mood, environment, or simply allow the environment around us to determine how we appear.

We have come a long way from the days of wrapping ourselves in the skins of animals. Welcome to the era of E-Textiles where we all can become creative designers expressing and defining our own inner desires, thoughts, and heart-felt feelings while truly wearing them on our sleeves!

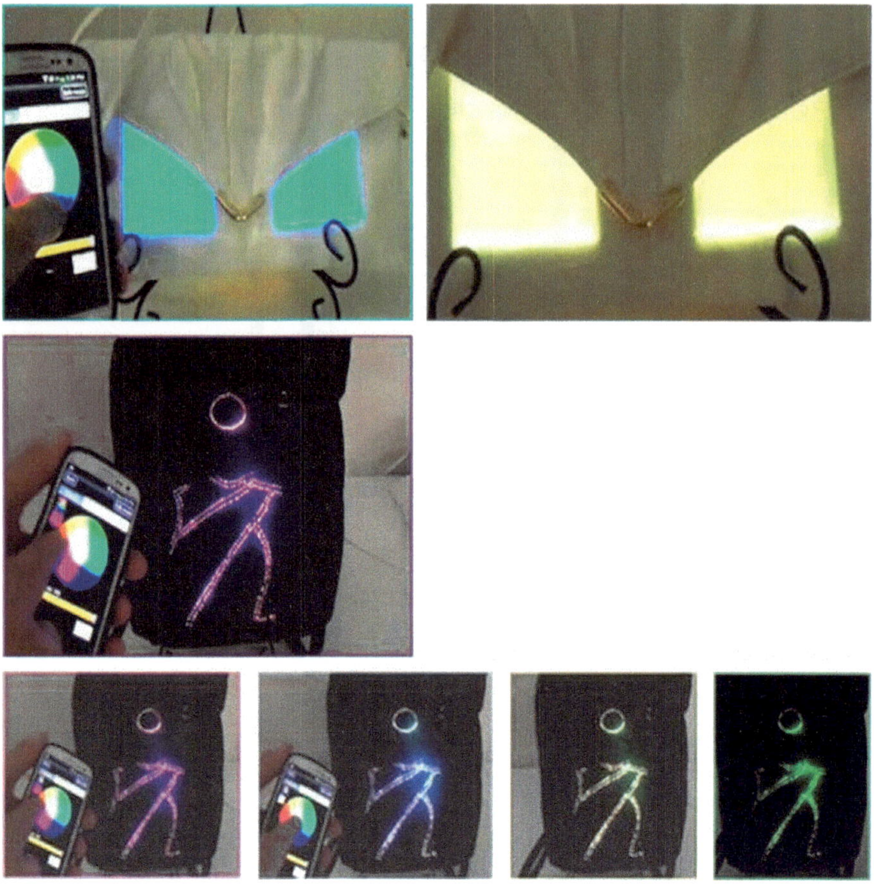

Fig. 9.17 Shows one method of controlling E-Textile surface color attributes by sliding a finger across a color wheel on a smart phone.
Photos provided by H. Lee Wainwright.

Source Links

http://www.adafruit.com/products/1643 (Neo-Pixel Ring-Rainbow Ring LEDs).
http://cutecircuit.com/ (Company examples of celebrity costumes using OLED arrays).
http://fashioningtech.com/ (Hair extensions) (Monitor Movements).
http://www.talk2myshirt.com/blog/archives/5617 (List of E-Textile companies engaging in research, prototyping, and applications).
http://etextilelounge.com/ (Conductive Thread evaluation comparisons with instruction videos).
http://www.ifmachines.com/ (Apparel Soft Switches).
http://www.ifmachines.com/products_faqs.html (Apparel electronics).
http://www.adafruit.com/category/92 (User-friendly accessories for building E-Textile projects).
http://www.plusea.at/?p=4203 (Textile sensor demonstrations).
http://www.visijax.com/about-us/our-technologies (Raise arms to activate turn signals in fabric).

http://www.rainbowwinters.com/videos.html (Photochromic, lenticular, and ambient driven color inks).

http://highlowtech.org/?cat=5 (MIT Media Lab Instructional books on Lilypad Arduino applications).

http://whisper.iat.sfu.ca/ (Whisper project 2007).

http://www.ife.ee.ethz.ch/research/groups/PlasticElectronics (Electronics Laboratory and wearable computing, Zurich, Switzerland—Thin film flexible electronics and sensors, smart textiles, and company spin offs).

http://subtela.hexagram.ca/index.html (Subtea project with scrolling jackets and flexible led sleeve digital displays, Concordia University, Canada).

http://www.ita.rwth-aachen.de/3-f-und-d/3-01-08-smart-textiles.html (Textile switching and functional control testing).

http://www.izm.fraunhofer.de/en/abteilungen/system_integrationinterconnectiontechnologies/arbeitsgebiete/elektronikintegrationinalternativematerialien/projects/integration_von_elektronikintextilien.html (Stretchable wiring and electronics + E-Textile testing and qualifying lab).

http://research.ocadu.ca/socialbody/projects (Social Body Projects at OCAD University, Toronto, Canada).

http://flextech.org/ (Flexible Display Conferences).

http://www.hleewainwright.com/resources/celebrity+examples.pdf (E-Textile Examples created by H. Wainwright).

https://www.smartfabricsconference.com/ (Conference covering Smart & E-Textile Research).

http://burningman.org/ (LED Lighting Event in Nevada Desert, USA).

http://onlinelibrary.wiley.com/doi/10.1002/elan.201000434/abstract (Goretex electrochemical E-textile detector for TNT and other explosives).

http://quantifiedself.com/2014/09/wrist-wearables-now/.

http://www.ncbi.nlm.nih.gov/pubmed/22163914/.

Thermal analysis for fiber identification and characterization

10

M.J. Smith
University of Minho, Braga, Portugal

10.1 Introduction

Textile fibers are essential components in a remarkable variety of commercial items including well-established traditional marketplaces in clothes, upholstery (office, household, and automotive), floor and wall coverings, cords and fasteners, but also increasingly in emerging applications in the fields of medicine, agriculture, and even as engineering materials in the construction industry. The chemical constitution of the fibers used in these applications is diverse and includes natural and manufactured fibers, with a growing tendency to combine different fibers in hybrid lattices, as warp-weft or entwined (plied) yarns, or as reinforcing structural elements in composite materials.

Fibers differ in their origin and chemical nature, their diameter, length, cross-section, surface contour, mechanical resistance, thermal stability, and color. All fibers are made up of macromolecules with their basic chemical unit being repeated hundreds or thousands of times. The structures of these units may be composed of homopolymers, block copolymers or branched copolymers with natural (plant, animal, or mineral) or synthetic origin.

As the chemical constitution of composite fibers becomes more complex and manufacturers of fabrics or finished products seek optimal operational characteristics in a long list of parameters that includes mechanical strength, dye fastness, easy-care or permanent press, perspiration wicking or thermal insulation, the response has been a broadening of the range of chemical and physical procedures to study fiber composition and characterize textile fabric performance. These studies have been driven by the need to understand structure–performance relationships, by the financial advantages that arise from providing better performance characteristics than competitors or even by the need of criminal investigators to establish a link between trace evidence found at a crime scene and a particular suspect. Whatever the motivation, the methods used to characterize the sample must provide reproducible and conclusive information that is only open to a single interpretation.

Thermal techniques including differential thermal analysis (DTA), thermogravimetric analysis (TGA), and thermomechanical analysis (TMA) are widely used in the characterization of polymers and fibers. The specific thermal behavior of a small amount of fiber sample may provide fingerprinting information that in itself is sufficient to identify a sample. These techniques may be complemented by other methods, including

Performance Testing of Textiles. http://dx.doi.org/10.1016/B978-0-08-100570-5.00011-6

optical or electron microscopy, infrared or fluorescence spectroscopy, or by traditional structural determination techniques to construct a complete profile of the sample.

There are many aspects of the behavior of a fiber sample that may be characterized by using thermal methods. These include the glass transition temperature, the melting temperature of principal components, percentage crystallinity, or the tendency to undergo cold crystallization. Often thermal processes associated with the release of surface water, process additives, behavior modifiers or solvent residues can complement the researcher's knowledge of the manufacturing process. Other events, such as phase transformations and thermal degradation of the bulk polymer or of specific additives, may also help complete the technical/behavioral profile of the fiber.

The objective of this chapter is to describe how thermal analysis can contribute to fiber identification and characterization. In order to achieve this objective it is necessary to understand how fibers are produced and how thermal history is impressed on these materials during production. The variety of commercial materials, manufacturing processes and additives used in the field of textile fibers renders a complete description impossible, however, an introduction to the principal aspects may be provided.

10.2 Formation and modification of fibers

Textile fibers from natural sources have been used for thousands of years but detailed chemical knowledge of their structure and understanding of their properties only became available during the 20th century. This understanding of natural fibers led to various key advances in textile chemistry. One of the most important advances was achieved when naturally occurring materials were transformed into fibers by extrusion, a process that rapidly evolved to large-scale production by melt spinning, dry spinning, and wet spinning. As the newly formed fibers exit the orifices of the spinneret head, the long chains of the polymer are aligned with the direction of flow and this process introduces orientation and crystallinity into the polymers. Crystallinity has a major impact on properties including fiber strength, flexibility, dye uptake, and color stability.

The discovery of Nylon constituted another key advance in textile chemistry. This synthetic polymer combines very attractive properties including considerable fiber strength, high elongation, elastic recovery, abrasion resistance, and flexibility. The packing density of these fibers can be adjusted to lower fabric air permeability, reduce water transfer, and improve water shedding. By density adjustment, the same fiber can provide high permeability with greater breathability and improved skin comfort. Nylon fibers are chemically resistant to cleaning agents and readily modified by chemical additives. Apart from its importance as a high performance fiber, Nylon signposted the way for future development of synthetic fibers and set a benchmark against which the performance of other fibers could be measured. From the middle of the 20th century to the present time the textile sector has undergone enormous changes as the raw materials transformed by the industry evolved from mostly natural to synthetic and hybrid sources and the variety of chemical and physical treatments increased dramatically to enable fine-tuning of the properties of commercial fibers.

Many of the treatments applied to alter the esthetic and functional performance also influence the thermal properties of fibers and it is therefore relevant to identify and comment on these processes.

A remarkable variety of chemicals are applied to fibers, yarns, and textiles with specific objectives that begin with the removal of natural impurities and processing residues to prepare the surface and permit penetration or adhesion of dyes or finishing agents. Chemical reagents may enhance comfort, provide low-maintenance finishes, retard combustion, or control bacterial agents. Antistatic finishes are used to reduce charge accumulation, the esthetic inconvenience of fabric cling, and the risks associated with the discharge of static electricity.

Not all fiber and fabric treatments are chemical in nature. Mechanical and heat treatments are combined in the drawing process applied to fibers as they emerge from the spinneret. The crystallinity of newly spun polymer is quite low and is increased by stretching and warming to allow individual polymer chains to move and form crystalline regions within the fibers. The process temperature, rate of flow, and stretching stress have a critical influence on the crystallinity of the product that results from the manufacturing process and, therefore, on the physical and chemical properties of the fiber. Mechanical crimping treatments are often used to change the appearance and increase the air space in the fabric bulk, and, therefore, also influence the way that the material conducts heat and contacts the skin. Singeing is applied to some fabrics to remove the free ends of fibers and render a smoother, more lustrous product. Heat-setting allows molecules of the fibers within the textile structure to move while the process temperature is maintained and permits retention of the "set" molecular orientation when the fiber cools in the fabric. Both these treatments alter the thermal properties of the chemical constituents of fibers and fabric.

All these chemical, thermal, or mechanical processes alter the surface or bulk behavior and as thermal and mechanical properties of polymers are sensitive to the presence of chemicals, treatment of fibers leaves a history that can be accessed and characterized by thermal analysis methods based on calorimetric, gravimetric, or mechanical processes.

A brief consultation of the patent literature (Sheperd, 1971; Sasaki et al., 1987; Ferenc et al., 2003) is sufficient to confirm that in textile chemistry the combination of various properties in a single item may provide a commercial advantage that results from differentiation between otherwise similar products. Various patents protect the production processes involved in the manufacture of nonwoven fiber mats that incorporate sources of fragrances applied to the fiber surface. More effective approaches (Ferenc et al., 2003) depart from the traditional strategy of surface application and describe fibers that contain hollow sections filled with high boiling point fragrances. Indicators that provide visual information regarding the reduction or exhaustion of the fragrance-releasing function can also be incorporated in this novel fiber structure. Fragrance-releasing dry or wet wipe fabric-based articles incorporating antimicrobial, antifungal, or insect-repellant preparations have been developed for commercial markets. While these fibers have extraordinary new properties, they, nevertheless, continue to retain the essential structural characteristics and aspects in their process history that are common to all manufactured fibers. As such, they are characterizable by thermal techniques and relevant information regarding the chemical and morphological state of these materials, which can be extracted to provide an identifiable profile.

10.3 Identification and characterization

Thermal analysis techniques can be used to characterize fibers or yarns composed of natural, manufactured, or synthetic materials. As described in the previous section, these materials can be chemically quite complex with microstructures that are partly determined by their chemical nature but are also greatly influenced by the treatments that they suffer during manufacture or during the fiber forming, finishing, or textile production stages. The range of properties obtained by appropriate processing of a selected synthetic fiber exceeds those obtained from the use of a wide variety of fibers with different chemical constitution (Jaffe et al., 1981). This underlines the remarkable impact and importance that processing parameters assume in the domain of fibers and textiles. Precise control over the conditions under which manufactured fibers are produced provides further practical advantages for this class of material relative to natural materials. In contrast, the properties of natural fibers from animal or plant sources are largely dominated by their chemical structure and therefore inevitably by the conditions under which they were produced.

In this section the objective is to discuss the origin of the thermal events observed in fibers and yarns and to explain how a detailed study of thermal events may lead to a better understanding of the relationship between process history, the characteristic properties of the fibers, and the structure of the fiber in question.

10.3.1 Thermal processes in fibers

The analysis of many classes of materials is undertaken with the purpose of discovering the chemical and physical history of the object. This is certainly the case in forensic analysis where substances collected from a crime scene can provide vital information that may incriminate suspected criminals. The exchange principle enunciated by Edmund Locard (Thornton, 1997) is also applicable to fibers in the sense that every stage in the synthesis or processing of the fibers or filaments, every thermal or mechanical treatment and every subsequent contact with chemicals leaves an indelible mark on the material. The challenge is to detect, characterize, and correctly interpret the information. Thermal analysis provides a set of tools that, effectively applied, can provide a wealth of information from minute samples.

The combination of chemical, thermal, and mechanical processes used in the production stages of continuous filament spinning leave a clear profile on the manufactured product. The formulation of the fiber-forming polymer pumped through the spinneret, the temperature of the cooling chamber, the nature of coagulant and the strain applied to the fiber, all contribute to the final characteristics of the filament. The fiber drawing and annealing steps also contribute changes to chain segment orientation and crystallinity and these are quantifiable using thermal analysis as glass transition and melting processes.

Twisting, air-jet processing and stuffer box or crimping processes are widely applied to introduce bulk into fiber by distortion, heating, and cooling. Transformation of yarns into garments, coverings or engineering textiles by matting, weaving, knitting or knotting subjects the components to further mechanical and thermal strain and contributes to the history of the textile. This residual mechanical strain in processed fibers, yarns, and textiles leaves a signature that can be read by thermomechanical methods of analysis.

The various stages of production that contribute to the thermal, mechanical, and chemical characteristics are particularly relevant to manufactured fibers. In contrast, the chemical nature of natural fiber structure plays a predominant role in determining the thermal characteristics. Natural fibers are usually characterized by a semicrystalline morphology and hydrophilic nature. The interaction of fibers with water is undoubtedly one of the most significant aspects of their behavior. In terms of thermophysical comfort, the interaction of fibers with water is important because the process of temperature management is centered on the transfer of water and the release of water vapor from the surface of skin, through garment fabrics, to the air layer of the environment in close contact with our bodies. This transfer determines the loss of heat in a hot environment and insulation from a cold environment by retention of a static layer of air. To provide comfort, a fabric has to be permeable to water and to be able to wick water away from the skin surface. Water may be imbibed or held between the fibers of a textile, between the fibers of a yarn, or be adsorbed onto the surface of fibers (Hatch, 1993). The ability to imbibe water is important, but fibers for use in garments must also release water to permit rapid drying after laundering. Water repellency is also important in applications in outer garments, but a balance between water repellency and permeability is essential to avoid accumulation of perspiration within the garment. The interaction of water with fibers is another identifiable parameter that influences the choice of fabrics for a given application. Many of the properties that affect the water–fiber interaction can be introduced or improved by appropriate chemical or physical treatment, but often it is the chemical composition of the fiber that determines specific interactions between the fiber and environmental humidity. Once again, thermal analysis is a useful tool that may be applied to evaluate the evaporation of imbibed or adsorbed water from a fiber, and may serve to distinguish between fibers. Both calorimetric and thermogravimetric analysis permit quantitative evaluation of the interaction of water with fibers and textiles.

The application of thermal analysis in the study of textile fibers is not restricted to their identification and characterization but may also be used to establish structure–performance relationships so that fiber chemistry can evolve and performance can be enhanced in future formulations and treatments.

10.3.2 Influence of experimental parameters in fiber analysis

From the preceding description of the structure of polymers in fibers, the mechanical and thermal stresses that are applied during manufacturing, texturing or textile fabric production, the additives that are deposited onto the surface during processing or with the purpose of introducing specific behavior in the final fiber, yarn or textile, it is clear that fibers from different sources must have different thermal behavior. All the stages of production have an influence, to a lesser or greater extent, on the thermal events that take place during thermal analysis. In general, these events are associated with the physical transformations of glass transition, crystallization, melting, the transformation of one solid phase into another or chemical transformations that take place at the fiber surface. Fibers that have water or residual processing solvents retained on their surface, or held between fibers, will release these species at characteristic temperatures. Chemical processes may include reactions between the fiber structure and the various coating components or between the fiber/coating and the local gas environment. Thermal degradation of the polymer structure will eventually take place as the bonds between the atoms in the structure begin to break and rearrange.

The interpretation of data from many analytical techniques is complicated by the existence of artifacts that can be introduced during sample collection and preparation. Thermal analysis methods also suffer from this problem and care must be taken, just as with the samples that are not of fiber origin.

10.3.2.1 Experimental aspects of DSC

The results of thermal analysis by differential scanning calorimetry (DSC) are influenced by various experimental parameters. Instruments designed and constructed by different manufacturers have distinct operating principles, but all instruments provide support for two crucibles of small volume in a single or a matched pair of furnaces that allow the temperature of the sample and a reference material to be subjected to a predefined temperature program. Various authors (Turi, 1981; Haines, 2002; Ehrenstein et al., 2004; Wunderlich, 2005; Brown and Gallacher, 2008; Menczel and Bruce Prime, 2009) have provided extensive descriptions of specific details of commercial instruments and therefore further aspects of equipment design will not be discussed in this chapter.

The chemical nature of the atmosphere that is in contact with the sample to be analyzed influences the thermal behavior of the sample and for this reason the DSC furnace is normally purged with a flow of an inert gas. This gas flow provides protection from atmospheric oxidation, removes decomposition products released from the sample and avoids ice formation when the furnace is operating at subzero temperatures. In certain cases it may be useful to study the reaction of the sample with other gases and in these circumstances the inert protective gas may be substituted, for example, by air as a reactive gas. The build-up of pressure inside the sample crucible during the heating program is avoided by punching one or more pin-holes in the lid of the crucible to allow gas exchange with the atmosphere inside the furnace.

Sample preparation is a critical step in thermal characterization of fibers. Samples of textile cloth, yarn or fiber are often studied by cutting the material into small pieces to facilitate their transfer to a small-volume crucible and establish a reproducible and efficient thermal contact with the interior surface of the crucible. The results shown in Fig. 10.1 (Steinmann et al., 2013) illustrate the importance of optimization of sample preparation. In this case significant differences arise from the use of clippings (or stripes), knots and bundles and from the choice of crucible volume. A clear advantage is observed in clipped fiber samples where this method of sample preparation permits an efficient contact with the interior surface of the crucible. The results obtained with clipped fibers are often further improved by using small-volume crucibles and deforming the lid after sealing the sample into the crucible so that it helps to compact the sample, establishing a good contact with the floor of the crucible.

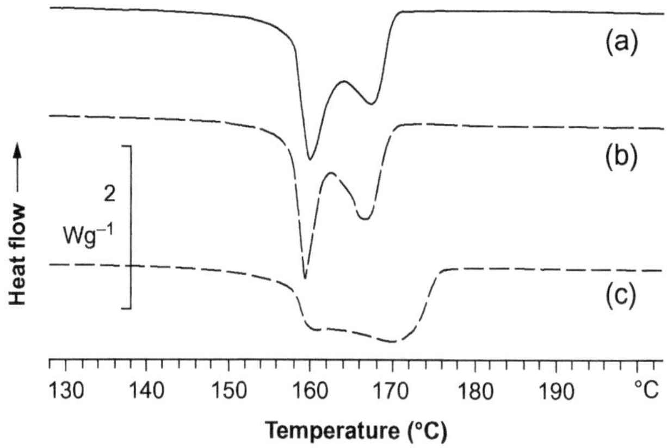

Fig. 10.1 Effect of sample presentation on thermal behavior of polypropylene fibers. Thermograms obtained with a 20 μL can at 5°C min^{-1} with (a) stripes (or short cut lengths), (b) knots, and (c) bundles.
Adapted from Steinmann, W., Walter, S., Beckers, M., Seide G., Gries, T., 2013. Thermal analysis of phase transitions and crystallization. In: Elkordy A.A. (Ed.), Polymeric Fibers, Applications of Calorimetry in a Wide Context—Differential Scanning Calorimetry, Isothermal Titration Calorimetry and Microcalorimetry. ISBN: 978-953-51-0947-1. InTech.: Rijeka, pp. 277–306. Available from: http://www.intechopen.com (accessed 7.7.15).

Some authors prefer to present samples as constrained or unconstrained lengths of coiled fibers. Many fibers shrink as they are heated and this physical change results in a corresponding alteration to the sample microstructure. The constraint of a fiber sample, by anchoring the ends or by winding onto a support, delays crystallization and crystal-perfecting and allows differences in the performance of similar materials to be detected (Menczel and Bruce Prime, 2009).

The mass of sample transferred to the crucible also requires optimization. The fusion of larger masses of sample requires more energy and generally it is advantageous, both in terms of peak width and resolution of multiple peaks, to use smaller sample masses. Of course the operator must use sufficient sample mass to observe small peaks that are caused by low intensity processes.

The quality of sample characterization by DSC experiments is significantly influenced by the choice of heating and cooling rates. In general, faster rates give rise to larger peaks, both in heating and cooling experiments. The results of analysis of polypropylene fibers (Menczel and Bruce Prime, 2009, Fig. 10.2) illustrate the importance of optimization of temperature program and demonstrate how peak size increases with heating rate. In this case peaks due to certain weak phase transitions are only detectable by using low heating rates.

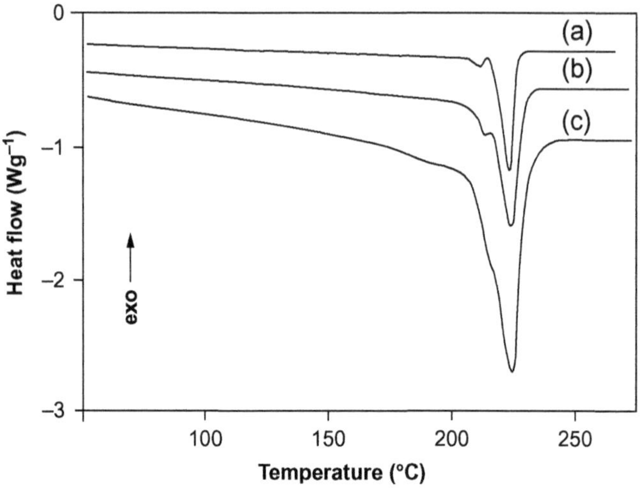

Fig. 10.2 Influence of heating rate on thermogram. Poly(butylene terephthalate) sample cooled from 250°C at 10°C min^{-1} then analyzed at indicated rates (a) 10°C min^{-1}, (b) 20°C min^{-1}, and (c) 40°C min^{-1}.
Adapted from Menczel, J.D., Bruce Prime, R., 2009. Thermal Analysis of Polymers: Fundamentals and Applications. John Wiley, New Jersey.

DSC experiments on fiber samples provide a means of characterizing materials by glass transition temperature, the extent and temperature of cold crystallization processes, the onset of crystalline phase melting and the temperature at which decomposition begins. Under certain circumstances the use of more than one heating cycle can provide additional information. During the first heating of the sample, aspects of thermal behavior that are due to the history imprinted on the sample during fabrication are erased. Philip (1972) described a simple procedure for identifying synthetic fibers based on thermal analysis of sample behavior during the second and subsequent heating cycles. This strategy effectively erases the as-fabricated thermal history of the fiber, replacing it with a known, internally comparable history that provides reproducible results for melting range, peak shape and heat of fusion.

10.3.2.2 Experimental aspects of TGA

The information obtained by the characterization of fibers by TGA can be particularly useful in that it provides access to a different viewpoint that often complements the data supplied by other thermal methods. While the results observed in DSC measurements for example, are greatly influenced by production steps that involve heating or stretching fiber-forming polymers to orientate the structural elements within the filament, the processes that give rise to loss of mass are much more closely related to the chemical nature of the fibers and the additives present within the structure and on the surface of the fibers. The reactions that originate the loss of mass recorded in TGA studies of fibers are generally caused by the evaporation of solvent, processing oil residues or water from within or between the fibers or filaments, or the chemical processes that occur between the fiber surface and the atmosphere inside the thermobalance (Treigienė and Musnickas, 2003; Moraes et al., 2010; Wielage et al., 1999). Various treatments with chemical components were described that are commonly used in processing or providing fibers, yarns and textiles with improved physical properties. These chemicals may alter the thermal stability of fiber-forming polymers and their presence may often be conveniently assessed by thermogravimetric evaluation (Ganan and Mondragon, 2002; Ouajai and Shanks, 2005; Riga et al., 1998; Chrissafis and Bikiaris, 2011).

The use of combined instrumental techniques is increasing as equipment design improves and interpretation of the data extracted becomes easier. The application of simultaneous DSC–TGA with subsequent analysis by advanced chromatographic and spectroscopic methods greatly facilitates interpretation of data. Transfer of evolved gas from the furnace to a gas chromatograph coupled to a mass spectrometer provides additional information that may help researchers to understand the roles that individual compounds play within the relatively complex matrices of synthetic and natural fibers. The information extracted from these experiments can be particularly relevant, for example in studies of flame resistant additives in military fabrics (Yip, 2014).

The application of TGA requires considerable attention to the preparation of samples if reliable and comparable results for degradation experiments are to be obtained. Thermogravimetric techniques are very sensitive to the surface-to-volume ratio (Jaffe et al., 1981, Fig. 10.3) and therefore consistent preparation of materials, and accurate reporting of experimental conditions including sample mass, heating rate and the nature of the furnace gas environment, are essential for authors to make sensible deductions from comparative studies. In general, small sample masses (5–10 mg), moderate heating rates (5–20°C min^{-1}) and very reproducible methods of preparation, particularly with fiber samples, are recommended.

10.3.2.3 Experimental aspects of TMA and DMA

The dimensional and mechanical stability of materials is often of critical relevance and may effectively determine their technical applicability. Synthetic polymers are processed at moderate to high temperatures to facilitate transformation into fibers with characteristics that are appropriate for specific applications. Residual "frozen-in" stress may result from the manufacturing process or from thermomechanical treatments that are applied to fibers as part of the finishing stage of production.

TMA is a technique that provides a measurement of sample dimensions while the specimen is subjected to a constant, usually small, mechanical stress (Turi, 1981; Haines, 2002; Ehrenstein et al., 2004; Wunderlich, 2005; Brown and Gallacher, 2008; Menczel and Bruce Prime, 2009). During this test the specimen is subjected to a controlled temperature program under a suitable atmosphere within an oven that is substantially larger than those found in DSC or TGA instruments. As the thermal inertia of the sample-support and position probe is relatively large, the heating rates used with TMA are lower. The thermal expansion coefficient can be determined and the manner in which this property alters can be followed as a function of temperature or time. The thermal expansion of a specimen changes as it approaches glass transition or melting

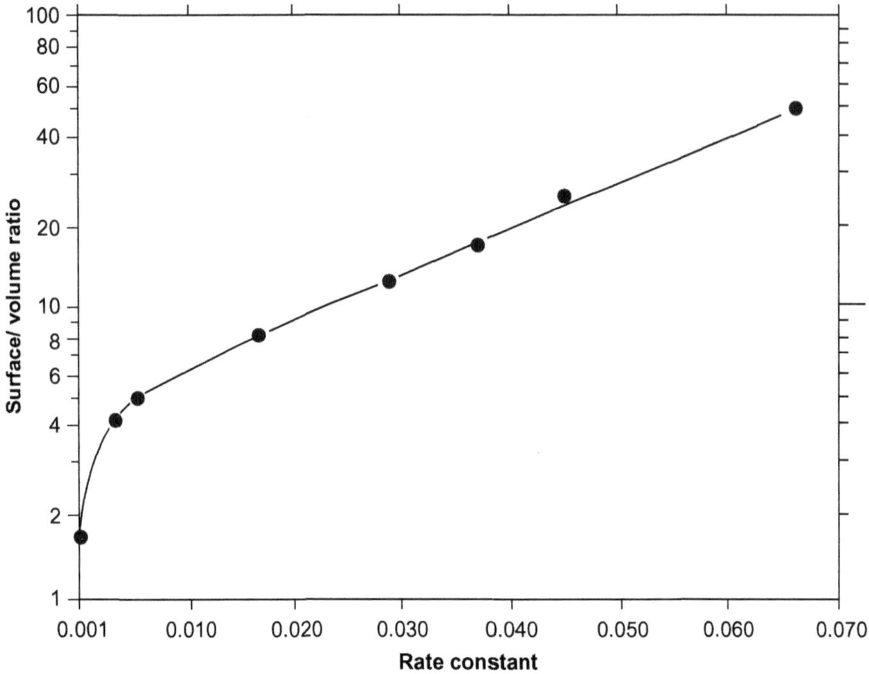

Fig. 10.3 Sensitivity of degradation rate constant to the surface-to-volume ratio of polypropylene yarn in air.
Adapted from Jaffe, M., Menczel, J.D., Bessey, W.E., 1981. Fibers. In: Turi, E. (Ed.), Thermal Characterization of Polymer Materials, second ed. Academic Press, New York, p. 1769.

temperatures. Materials that have residual stress generally show dimensional changes that can be followed as a function of temperature. As the temperature increases, the molecules in the structure begin to move and the "frozen-in" stress begins to relax. Although various compression, penetration and volumetric TMA probes are commonly used to evaluate samples fabricated from different materials, the tension probe is most appropriate for samples like fibers or yarns that are not self-supporting. In the TMA experiment the fiber is subjected to a constant force and a proximity sensor registers the movement of the probe. In common with other thermal analysis techniques, the response of processed fibers is determined by the sample history and the experimental parameters must be carefully controlled if consistent and accurate results are to be obtained. Sample preparation, choice of heating rate and control of test chamber atmosphere must be given due attention.

The TMA technique provides a means by which the behavior of a sample may be characterized in response to specific experimental circumstances and allows a detailed "fingerprint" to be recorded as small sample history differences give rise to different dimension-temperature behaviors. Properties that are of immediate relevance to end users, including linear thermal expansion coefficient, thermal shrinkage and the shrinkage force, may be obtained by direct measurement from TMA characterization.

In experiments where the starting temperature of the fiber is below the glass transition temperature, the first region of the zero load shrinkage thermogram shows linear expansion up to T_g. Under certain circumstances irreversible shrinkage may be observed in this region, for example as a result of liberation of solvent or moisture adsorbed during fiber production. The value of dS_T/dT, the first derivative of sample shrinkage, is constant and this provides the coefficient of thermal expansion of the sample (Jaffe et al., 1981). When the T_g of the sample is reached, the fiber suffers a rapid irreversible shrinkage as the polymer chains in the amorphous fraction of the fiber relax. The change in behavior at this temperature provides a means by which T_g can be estimated. Between T_g and the melting temperature, T_m, shrinkage takes place as a result of processes that occur in the sample bulk, including polymer chain reorganization, chain folding, recrystallization, and crystal perfecting. In this region of the shrinkage thermogram, the behavior once again reflects the nature of the process history of the fiber. A further change in the thermogram is observed as the sample approaches the melting temperature of the crystalline domains. The crystallites in the fiber structure begin to transform, enter the amorphous phase and polymer chains leave the structured regions of the fiber. The peak observed at this point in the dS_T/dT curve is useful for identifying the melting point of the polymer crystallites and confirms the onset of the melting peak obtained by DSC.

All the processes involved during the heating of the fiber are very dependent on the thermomechanical treatment that the fiber suffered during manufacture and for this reason TMA is often considered to be the most useful thermal technique for obtaining information about the production procedure applied to a sample fiber.

In several respects, dynamic mechanical analysis (DMA) is similar to TMA and is often used to study the phase behavior of polymer and polymer composites. An important difference between the two techniques is that in DMA an oscillating stress is applied to the specimen and the instrument has the additional ability to resolve specimen deformation into in-phase and out-of-phase components such that E', E'' and $\tan\delta$ can be determined. Many instruments are capable of applying mechanical stress over a range of operating frequencies.

10.3.3 Examples of thermal behavior of fibers

In view of the restrictions in space, it is impractical to review the immense volume of literature that describes the application of thermal analysis methods to textile fibers, yarns, and fabrics. The objective of this section is therefore to describe examples of thermal processes in textile materials that have been selected to illustrate the behavior observed in specific samples. Comparison with these examples may help to identify and interpret similar processes and understand the performance of other materials.

In recent years poly(ethylene terephthalate) (PET) microfibers (De Clerck et al., 2003) have attracted significant attention as a result of their superior drape, hand, improved water repellence, and breathability. Unfortunately, these advantages are also accompanied by difficulties in achieving a uniform color build-up in dyeing. Heat-set processes are applied to many fibers, yarns, and fabrics to improve their dimensional stability and this involves exposure of the material to an annealing treatment.

The application of color to both fiber and microfiber PET typically takes place using disperse dyes under pressure at temperatures close to 130°C. These conditions have been chosen to present the polymer structure of the fibers in an amorphous state because efficient diffusion of dye into fibers only occurs when the polymer strands become mobile. This dye application temperature is below the heat-set temperature and therefore does not alter the morphology of the material.

It is often difficult to study broad and weak glass transitions with conventional DSC and in these cases the technique known as modulated temperature DSC (MTDSC) technique provides greater sensitivity. MTDSC is an important innovation that was introduced by Reading and coworkers (Seferis et al., 1992; Gill et al., 1993) and has been developed by various equipment manufacturers. In this technique a controlled periodic temperature modulation is superimposed on the sample heating or cooling program and it becomes possible to separate total heat flow into reversing and nonreversing components (Haines, 2002; Wunderlich, 2005; Brown and Gallacher, 2008; Menczel and Bruce Prime, 2009). Where thermal events are superimposed, a complication that often arises in thermal analysis of many materials, it may be possible to separate overlapping transitions and reveal effects that are not visible using conventional DSC instruments. Fig. 10.4 demonstrates the successful application of this technique. In Fig. 10.5 the heat capacity signal, obtained by MDSC, allows determination of the glass transition of the microfiber sample.

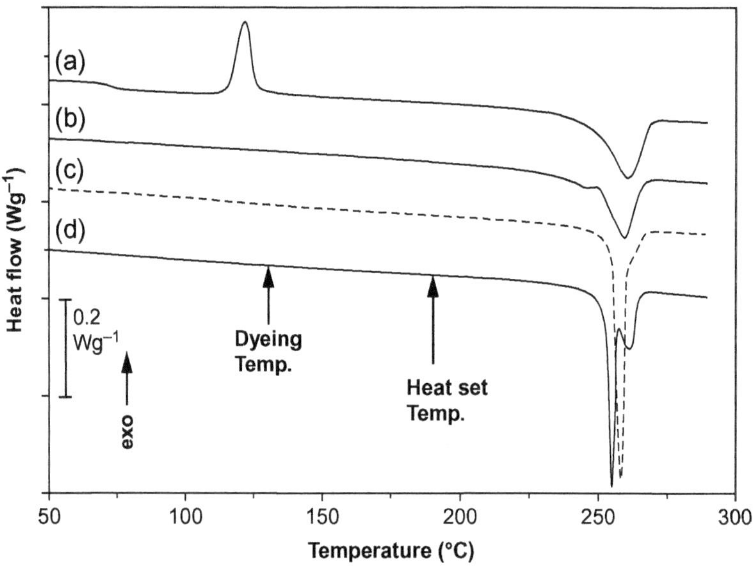

Fig. 10.4 Conventional DSC thermogram of (a) amorphous bulk PET, (b) semicrystalline bulk PET, (c) microfiber PET, and (d) conventional PET fiber.
Adapted from De Clerck, K., Rahier, H., Van Mele, B., Kiekens, P., 2003. Thermal properties relevant to the processing of PET fibers. J. Appl. Polym. Sci. 89, 3840–3849.

The variation of dCp/dT with sample temperature permits a more precise identification of the onset of even quite weak glass transitions. The thermal characterization of PET microfibers is particularly relevant because the processing of these materials results in a high molecular orientation, a high crystallinity, a substantial increase in the onset of T_g and a broadening of the temperature range over which the transition takes place. Using the information obtained from this study it is clear that the empirically-determined dyeing temperature must indeed be close to 130°C in order to exceed the fiber T_g and allow appropriate dye build-up to be achieved. Thorough characterization of fibers can clarify the complex relationship between dye application, heat-setting and the thermal properties of commercial fibers.

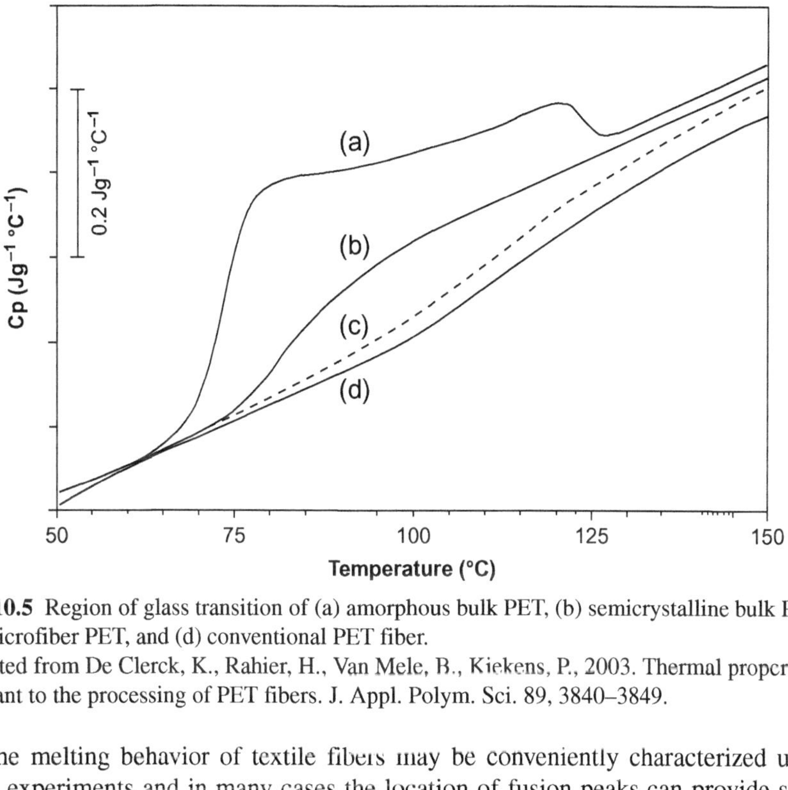

Fig. 10.5 Region of glass transition of (a) amorphous bulk PET, (b) semicrystalline bulk PET, (c) microfiber PET, and (d) conventional PET fiber.
Adapted from De Clerck, K., Rahier, H., Van Mele, B., Kiekens, P., 2003. Thermal properties relevant to the processing of PET fibers. J. Appl. Polym. Sci. 89, 3840–3849.

The melting behavior of textile fibers may be conveniently characterized using DSC experiments and in many cases the location of fusion peaks can provide sufficient information to identify the polymer sample. This process is rarely simple where the samples are manufactured fibers (Lacey et al., 2006). Normally fibers are semicrystalline polymers and during heating the transfer of material from the crystalline component to surrounding amorphous regions takes place over a range of temperature and both the identification of the melting temperature and the evaluation of the heat of fusion are processes that require special care. Neither the extrapolated onset of the fusion peak, nor the peak temperature, corresponds to the melting temperature of such a system. The melting point is conventionally taken as the temperature at which the most perfect crystallites melt at the end of the melting peak and a sigmoidal baseline should be used to determine the heat of fusion (Menczel and Bruce Prime, 2009). DSC melting curves obtained with textile fibers report the heat flow during a nonequilibrium process. Melting curves acquired at different heating rates are not superimposable because processes of polymer chain rearrangement, recrystallization, and superheating occur. These processes are time-dependent. Polymer crystals subjected to heating are able to undergo perfecting as chains in the immediate neighborhood of a small crystallite within lamellae become aligned and cause crystallite growth. An increase in the heating rate used to characterize the sample causes a decrease in the onset of fusion and often alters the shape of the melting peak as the faster heating rate is not sufficient for the polymer chain reorganization to take place. Recrystallization occurs during melting as polymer chain segments are released and their mobility allows them to align and form crystals. This is an exothermic process and can often be observed as an exothermic peak between two endotherms, when the heating rate permits. Again, alteration of the experimental heating rate causes a change in the thermogram. When recrystallization is the cause of the thermal effect observed, the height of the first melting peak becomes smaller relative to the second peak as the heating rate is increased. This reflects the inability of the polymer to recrystallize within the time available before the melting temperature of the polymer crystallite is reached. Superheating is the term used to designate a situation in which the melting temperature of a sample increases as the heating rate is increased. It is caused by the rate of heat supply to the sample being faster than the rate of crystal melting.

The example of the thermal characterization of drawn fibers of Nylon 6, prepared for textile fabric production, can provide illustration of complex melting behavior (Kubokawa and Hatakeyama, 2002) in semicrystalline samples presented as oriented fibers. In the experiments described by these authors the curve identified as (a) in Fig. 10.6 was a Nylon fiber sample cut into short lengths and packed into an aluminum can. Sample (b) was prepared by transferring a 4 mm disc cut from the fabric into a can, and sample (c) was prepared by winding warp and weft fibers from unraveled fabric onto a disc to provide a restrained fiber. The results were interpreted in terms of partial melting and reorganization of crystallites during heating, giving rise to a triple endotherm in curve (a) of Fig. 10.6. The presentation of the same sample as a fabric or constrained fibers results in alterations to the resolution, the intensity of peaks and to the temperatures at which events take place. This example calls attention to the extent and complexity of the effects of sample preparation and to the importance of appropriate and consistent sample presentation. At the same time, the usefulness of the DSC technique as a manner of highlighting very small differences between samples can also be appreciated.

The fusion of fibers is often further complicated by melting of different crystalline forms, or by crystal to crystal transitions that occur between different forms of the same polymer. The heat treatment of fibers has already been described, and the process of annealing can give rise to thermograms that assume quite different appearances as different thermal treatments are imposed on the sample. Fig. 10.7 (Menczel and Bruce Prime, 2009) illustrates the study of samples of low-density poly(ethylene) that were heated to above melting, cooled, annealed at various temperatures, quenched, and reheated at 10°C min^{-1}. The small peak is caused by the fusion of crystallites with melting temperatures lower than T_{ann} and recrystallizing to form structures with crystal perfection determined by T_{ann}. This example demonstrates the need for care in interpreting the thermographs of annealed and heat-treated polymer systems.

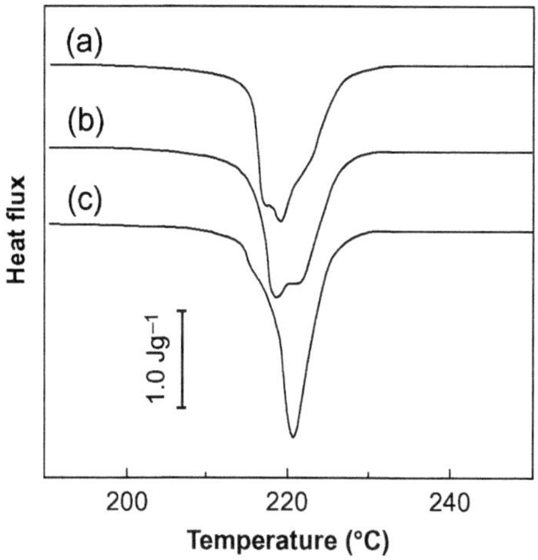

Fig. 10.6 Thermograms of heating curves of Nylon 6 fabrics illustrating effect of fiber presentation on multiple melting peaks. (a) short cut fibers, (b) textile fabric disk, (c) restrained warp and weft fibers.
Adapted from Kubokawa, H., Hatakeyama, T., 2002. Melting behavior of Nylon 6 fibers in textiles. J. Therm. Anal. Calorim. 70, 723–732.

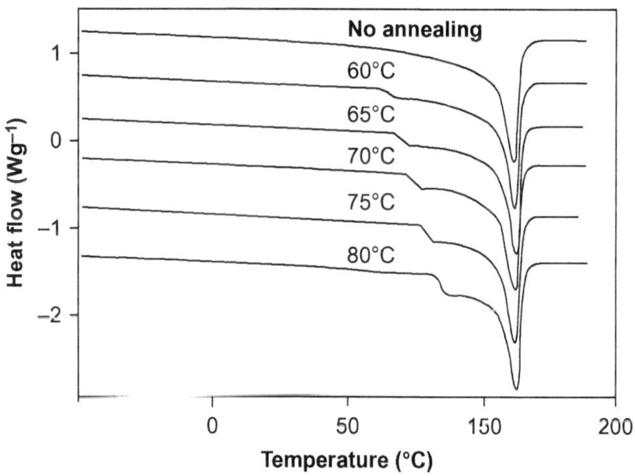

Fig. 10.7 Melting curves of low-density poly(ethylene) after annealing treatments at the temperature indicated on the curves.
Adapted from Menczel, J.D., Bruce Prime, R., 2009. Thermal Analysis of Polymers: Fundamentals and Applications. John Wiley, New Jersey.

Thermogravimetric analysis provides information that is complementary to that obtained by DSC. The technique is particularly useful where changes of mass occur as a result of drying, volatilization, chemical reaction, and degradation. In thermogravimetric studies, the relevant thermal processes often begin between 100°C and 150°C where adsorbed water or other volatiles are released, followed by partial loss of mass where the fiber suffers structural alteration with loss of polymer fragments, with further major loss of mass caused by subsequent degradation and the formation of a residual mass of charred material remaining after degradation. Typical applications of TGA in polymer chemistry include evaluation of thermal stability, or estimation of service lifetime, under well-defined conditions of temperature and in known reactive or inert gas environments. Many natural and manufactured fibers support combustion and in certain applications, for example when users are exposed to fire-risk environments, it is important to be able to compare the performance of different fibers and the efficiency of flame-resistant treatments. This specific test requirement is relevant for clothing of military and firefighting personnel and in textiles for bedlinen and drape materials. Performance parameters including the ease of flame extinction, the amount of heat released during combustion and the temperature at which degradation begins may be quantified using thermal methods. The results illustrated in Fig. 10.8 (Yip, 2014) show the derivative weight as a function of temperature and demonstrate how differences between fibers may be highlighted using this technique. While these methods are not representative of a real fire scenario, they do provide comparable test methods that may order fibers and treatments by merit and quantify the loss of effectiveness of treatment through laundering.

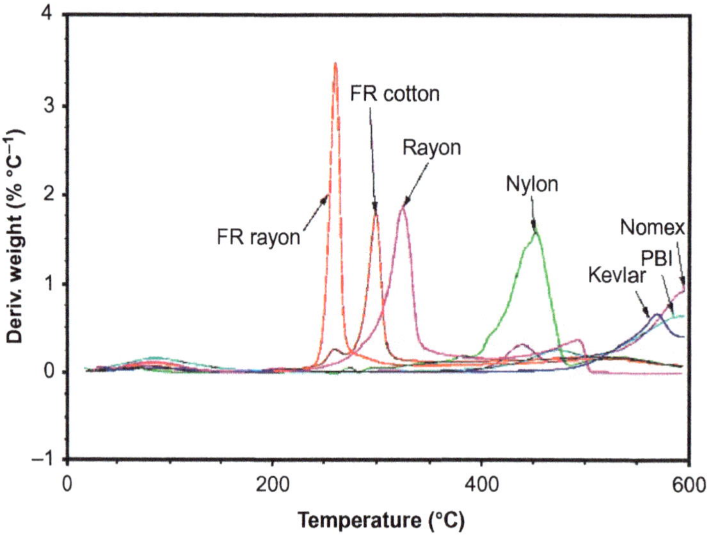

Fig. 10.8 TGA as a screening method for fire-resistance fibers. Derivative weight is shown as a function of temperature curves, under air. Fibers are fire-resistant (FR) Rayon, FR cotton, Rayon, Nylon, Kevlar, poly(benzimidazole), and Nomex.
Adapted from Yip, P.W., 2014. Analysis of two methods for characterization of flame resistant military fabrics and commercial textile fibers: simultaneous DSC-TGA and pyrolysis GCMS. Technical report NATICK/TR-14/008. US Army Natick Soldier Research, Development and Engineering Center, Natick, MA.

Thermogravimetric characterization does not just provide quantitative information regarding the relative stability of manufactured fibers (Carrillo et al., 2004), and establish a relationship between molecular orientation arising in the spinning process and the thermal stability of the finished fiber, but may also quantify water uptake (Jaffe et al., 2003) or desorption and compare the intensity of water–fiber interactions. The evaluation of weight loss as a function of temperature can provide a convenient means of comparing the thermal release of components from fibers treated with essential oils and nanocapsules containing fragrances (Liu et al., 2015) or the changes of degradation characteristics of fabrics treated with mordants and dye molecules (Ibrahim et al., 2011).

The instrumental methods of thermomechanical or DMA are particularly well adapted to the study of synthetic and natural fibers and stress relaxation processes in fibers, yarns, and textiles. The superior mechanical strength, biocompatibility, and biodegradability of natural silk fibers led to the development of lower cost synthetic fibers from regenerated silk fibroin (RSF). Similar performance characteristics have been achieved by wet-spinning RSF into a coagulation bath with fiber drawing and steam annealing (Zhou et al., 2009). DMA has been used to characterize RSF films and correlate the results of dynamic tan δ studies with structural changes that occur during crystallization, thermal dehydration, and drawn fiber orientation (Yuan et al., 2010). In studies of various samples produced by natural spinning, forced-reeling, annealing and post-drawing fibers, DMA was used to evaluate the behavior of silk samples of different origins and assess the effect of fiber production procedures on the resulting fibers (Mortimer et al., 2015; Fig. 10.9). The authors concluded that DMA characterization allowed the quality of various silk fibers to be compared and permitted evaluation of post-spin treatments.

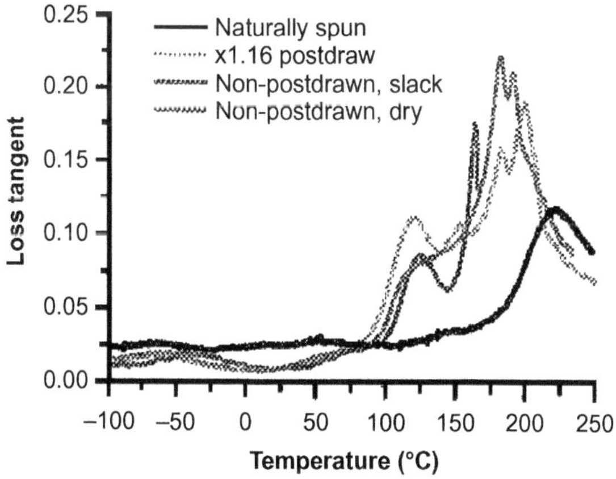

Fig. 10.9 Loss tangent profiles of force-reeled silk fibers (naturally spun, postdrawn sample, slack-stored and not-postdrawn and dry-stored and not-postdrawn).
Adapted from Mortimer, B., Guan, J., Holland, C., Porter, D., Vollrath, F., 2015. Linking naturally and unnaturally spun silks through the forced reeling of *Bombyx mori*. Acta Biomater. 11, 247–255.

The process of dyeing is an operation of crucial importance to the commercial value of textile fibers. In many cases water is chosen as dye solvent, for dye transportation and also as an agent to swell or plasticize the fiber during dyeing. The process temperature is of critical importance in dyeing because dye penetration and diffusion occur with more uniform color uptake at temperatures greater than the T_g of the fiber. Various thermal analysis techniques, including DSC, TMA, and DTA can be used to evaluate T_g of fibers, however the possibility of measuring T_g under wet conditions offers significant advantages. Aitken and coworkers (Aitken et al., 1991) modified the furnace and sample clamp assembly of a DMA to permit the study of scoured acrylic fibers to be carried out under controlled wet conditions. Comparison of the T_g of dry fibers using DSC and DMA confirmed that a peak was observed at 92°C in the $\tan\delta$ versus temperature behavior, a value substantially different from that of 75°C with wet fiber. These results were interpreted as confirmation of the plasticizing action of water on acrylic fibers and were in agreement with previous studies of acrylic polymers (Hor et al., 1988).

10.4 Evolution of novel fiber materials

The textile industry is a domain sustained and driven by the development of new fibers, treatments and production strategies. The integration of developments from the field of nanotechnology into this dynamic industry provides exciting opportunities for discontinuous evolution, in the sense that the combination of existing materials with novel functionality is expected to introduce properties, some of which are illustrated in Fig. 10.10, that are quite different from those available through traditional means (Soane et al., 2010; Sawhney et al., 2015).

Traditional surface treatments (Hatch, 1993) are often limited because laundering, or even normal use of the treated garment, may cause removal of the active agents. The use of chemically bound reagents on the fiber may also have a negative influence if restrictions at the binding site alter the mechanical behavior of the fiber. Modification at air-fabric and fluid-fabric interfaces is of particular relevance to textile chemistry because many performance characteristics are critically dependent on the behavior of this region.

Attributes of liquid repellency and stain resistance (Gulrajani, 2006) are properties that are particularly attractive in work attire and uniforms. Fiber surface modification by nanowhiskers introduces advantages in stain management without alteration to breathability, touch softness, and user comfort. Multifunctional fluorocarbon molecules form bonds with cellulosic and proteinous fibers and orientate side-chains to produce a robust hydrophobic and oleophobic finish that shows good wash-fastness (Linford et al., 2005). The application of suitable chemical components (Barthlott, 1996) to fiber surfaces introduces self-cleaning functions that mimic those of leaf surfaces (Neinhuis and Barthlott, 1997) by providing nanoscale elevations and depressions on the modified surface.

Integrated UV protection of textiles is becoming increasingly important as the detrimental effects of radiation are recognized and textiles are used as a barrier during outdoor activity. Inorganic nanoparticles (15–70 nm) are an effective substitute for

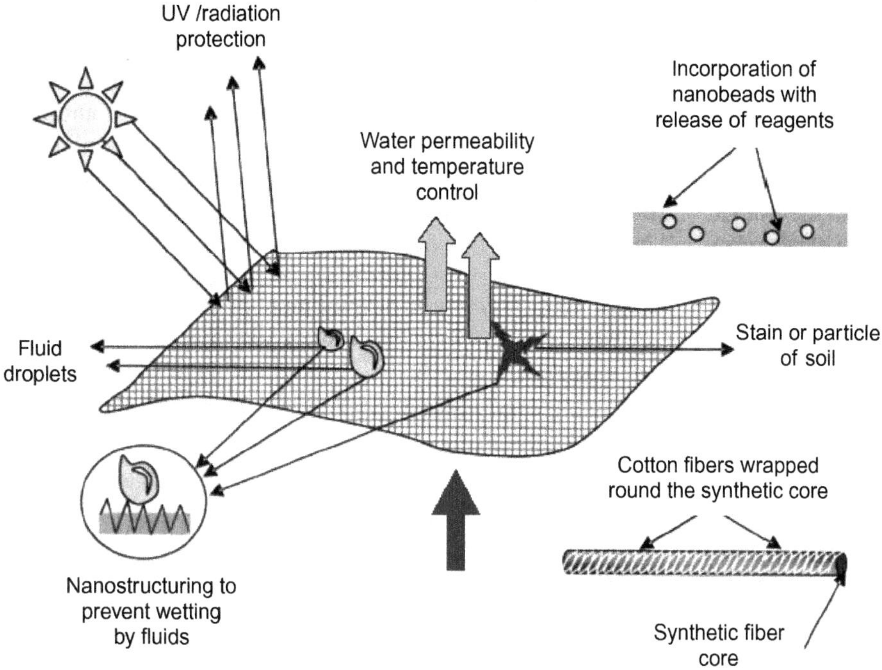

Fig. 10.10 Contribution of nanotechnology to fiber functionality.
Adapted from Sawhney, A.P.S., Condon, B., Singh, K.V., Pang, S.S., Li, G., Hui, D., 2015. Modern applications of nanotechnology in textiles. Text. Res. J. 78, 731–739.

organic UV absorber molecules (Riva et al., 2006; Ibrahim and El-Zairy, 2009) as their effectiveness is not reduced by radiation destruction and they are more resistant to leaching during laundering. Substantial performance improvements are expected through the use of nanoparticles (Tsuzuki and Wang, 2010) and multifunctional coatings that have been applied to a wide variety of fibers (Beringer and Hofer, 2004).

Core-wrapping (Collier, 1995) is the process through which a permanently attached sheath of carbohydrate is applied to a synthetic fiber to combine the high strength and durability (Hakansson et al., 2014) of the core with the desirable properties of natural hydrophilic wraps. The reduction of static charge accumulation in synthetic fibers is traditionally accomplished by the application of antistatic finishes. In core-wrapped fibers the hydrophilic nature of the carbohydrate sheath provides more effective charge dispersion than surface finishes that are progressively removed by laundering.

A nanotechnological-based approach can provide advantages of controlled agent-release. Nanobeads or polymeric encapsulators (Tree-Udom et al., 2011; Soane et al., 2010) may be permanently attached to fabrics with textile-reactive groups so that the payload, carried and protected (Tree-Udom et al., 2011) by the capsule, may be released in a controlled and prolonged fashion (Rodrigues et al., 2008). This payload may be a bioactive, antimicrobial or antifungal agent, a pharmaceutical drug, sunblock reagent, dye or pigment, scent or fragrance, insect repellent, fire retardant, fabric softener, phase change agent, or other substance designed to provide the textile with special properties.

Although these materials, incorporating nanotechnological functionality, use innovative strategies in their manufacture or during finishing, fibers, yarns, and textiles still contain base components that, by their composition and chemical nature, are entirely amenable to thermal characterization. Thermal methods therefore continue to provide very relevant information about process variables, permit assessment of fiber/textile properties, and quantify the performance advantages that have been achieved.

10.5 Conclusion

Thermal methods have evolved significantly over recent years as new developments have brought improved reproducibility, increased precision of measurement and improved convenience of use through the continuous evolution of instrumental design. The fundamental transformations that are registered as the sample are subjected to the temperature program used in thermal analysis, however, they have not changed and continue to provide relevant information regarding the chemical nature and thermomechanical behavior of the sample.

Fibers and textiles continue to evolve to take advantage of the opportunities that are offered by integrating nanotechnology into traditional and novel components. Materials that incorporate biodegradability or biocompatibility are opening new markets for a wide range of products and the information provided by thermal analysis allows researchers to identify and characterize textile fibers by measuring physical properties including the glass transition temperature, crystallization, melting range, thermal stability and thermal-oxidative stability.

The future importance of this group of techniques lies in its capacity to identify and rationalize the structural and morphological aspects of a wide range of fibers and to understand how production strategy can lead to the observed performance. The recognition of structure–performance relationships will certainly lead to alterations that provide access to improved technical characteristics of a great variety of specialty fiber products.

References

Aitken, D., Burkinshaw, S.M., Cox, R., Catherall, J., Litchfield, R.E., Price, D.M., Todd, N.G., 1991. Determination of the T_g of wet acrylic fibers using DMA. J. Appl. Polym. Sci. 47, 263–269.

Barthlott, W., WO 04123, Bonn, Germany, 1996 Eur Pat 772514 BI, 15 February 1996.

Beringer, J., Hofer, D., 2004. Nanotechnology and its application. Melliand Int. 10, 295–296.

Brown, M.E., Gallacher, P.K., 2008. Handbook of Thermal Analysis and Calorimetry, vol. 5. Elsevier, Oxford.

Carrillo, F., Colom, X., Sunol, J.J., Saurina, J., 2004. Structural FTIR analysis and thermal characterisation of lyocell and viscose-type fibres. Eur. Polym. J. 40, 2229–2234.

Chrissafis, K., Bikiaris, D., 2011. Can nanoparticles really enhance thermal stability of polymers. Thermochim. Acta 523, 1–24.

Collier, J.R., Collier, B.J., 1995. Process of Making Sheath/Core Composite Products. Lousiana State University, Assignee US Patent 5,387,383, 7 February.

De Clerck, K., Rahier, H., Van Mele, B., Kiekens, P., 2003. Thermal Properties Relevant to the Processing of PET Fibers. J. Appl. Polym. Sci. 89, 3840–3849.

Ehrenstein, G.W., Riedel, G., Trawiel, P. (Eds.), 2004. Thermal Analysis of Plastics: Theory and Practice. Carl Hanser Verlag, Munchen.

Ferenc, D., Novas, E.S., D'Áscanio, L.O., 2003. Method of Making Fragrance Containing Fiber. Assignee: International Flavors & Fragrances, Inc. US Patent US 6517759 B1, published 11 February.

Ganan, P., Mondragon, I., 2002. Surface modification of fique fibers. Effects on their physico-mechanical properties. Polym. Compos. 23, 382–394.

Gill, P.S., Sauerbrunn, S.R., Reading, M., 1993. J. Therm. Anal. 40, 931–939.

Gulrajani, M.L., 2006. Nano finishes. Indian J. Fibre Text. Res. 31, 187–201.

Haines, P.J. (Ed.), 2002. Principles of Thermal Analysis and Calorimetry. Royal Society of Chemistry, Cambridge, United Kingdom.

Hakansson, K.M.O., Andreas, B.F., Lundell, F., YU, S., Krwka, C., Roth, S.V., Santoro, G., Kvik, M., Wittberg, L.P., Wagberg, L., Soderberg, L.D., 2014. Hydrodynamic alignment and assembly of nanofibrils in strong cellulose filaments. Nat. Commun, 1–10.

Hatch, K., 1993. Textile Science. West Publishing Company, New York.

Hor, T., Khang, H., Shimuzu, T., Zollinger, H., 1988. Text. Res. J. 58, 227–269.

Ibrahim, N.A., El-Zairy, E.M.R., 2009. Union disperse printing and UV-protecting of wool/polyester blend using a reactive b-cyclodextrin. Carbohydr. Polym. 76, 244–249.

Ibrahim, S.F., El-Amoudy, E.S., Shady, K.E., 2011. Thermal analysis and characterization of some cellulosic fabrics dyed by a new natural dye and mordanted with different mordants. Int. J. Chem. 3, 40–54.

Jaffe, M., Menczel, J.D., Bessey, W.E., 1981. Fibers. In: Turi, E. (Ed.), Thermal Characterization of Polymer Materials, second ed. Academic Press, New York, p. 1769.

Jaffe, M., Ophir, Z., Pai, V., 2003. Biorelevant characterization of biopolymers. Thermochim. Acta 396, 141–152.

Kubokawa, H., Hatakeyama, T., 2002. Melting behavior of nylon 6 fibers in textiles. J. Therm. Anal. Calorim. 70, 723–732.

Lacey, M., Price, D.M., Reading, M., 2006. Theory and practice of modulated temperature differential scanning calorimetry. In: Reading, M., Hourston, D. (Eds.), Modulated Temperature Differential Scanning Calorimetry. Springer, Dordrecht, The Netherlands (Chapter 1).

Linford, M.R., Soane, D.S., Offord, D.A. and Ware, W. Jr., Durable finishes for textiles. Assignee: Nano-Tex Llc. US Patent 6,872,424, 29 March 2005.

Liu, C., Liang, B., Shi, G., Li, Z., Zheng, X., Huang, Y., Lin, L., 2015. Preparation and characteristics of nanocapsules containing essential oil for textile application. Flavour Fragance J. 30, 295–301.

Menczel, J.D., Bruce Prime, R., 2009. Thermal Analysis of Polymers: Fundamentals and Applications. John Wiley, New Jersey.

Moraes, M.A., Nogueira, G.M., Weska, R.F., Beppu, M.M., 2010. Preparation and characterization of insoluble silk fibroin/chitosan blend films. Polymers 2, 719–727.

Mortimer, B., Guan, J., Holland, C., Porter, D., Vollrath, F., 2015. Linking naturally and unnaturally spun silks through the forced reeling of *Bombyx mori*. Acta Biomater. 11, 247–255.

Neinhuis, C., Barthlott, W., 1997. Characterization and distribution of water-repellent, self-cleaning plant surfaces. Ann. Bot. 79, 667.

Ouajai, S., Shanks, R.A., 2005. Composition, structure and thermal degradation of hemp celluloseafter chemical treatments. Polym. Degrad. Stab. 89, 327–335.

Philip, W.M.S., 1972. The use of differential scanning calorimetry in the identification of synthetic fibers. J. Forensic Sci. 17, 132–140.

Riga, A., Collins, R., Mlachak, G., 1998. Oxidative behavior of polymers by thermogravimetric analysis, differential thermal analysis and pressure differential scanning calorimetry. Thermochim. Acta 324, 135–149.

Riva, A., Algaba, I.M., Pepió, M., 2006. Action of a finishing product in the improvement of the ultraviolet protection provided by cotton fabrics. Modelisation of the effect. Cellulose 13, 697–704.

Rodrigues, S.N., Fernandes, I., Martins, I.M., Mata, V.G., Barreiro, F., Rodrigues, A.E., 2008. Microencapsulation of limonene for textile application. Ind. Eng. Chem. Res. 47, 4142–4147.

Sasaki, M., Yoshida, J., Shimizu, Y., Akazawaa, K., Kishita, E., Hirata, K., 1987. Fragrant fiber. Assignee: Mitsubishi Rayon Company Ltd. US patent US 4713291 A, published 15 December 1987.

Sawhney, A.P.S., Condon, B., Singh, K.V., Pang, S.S., Li, G., Hui, D., 2015. Modern applications of nanotechnology in textiles. Text. Res. J. 78, 731–739.

Seferis, J.C., Salin, I.M., Gill, P.S., Reading, M., 1992. Proc. Acad. Greece 67, 311.

Sheperd, T.H. and Gould, F.E. Entrapped essences in dry composite fiber base products giving a strong fragrance when wet in water. Assignee: Nat Patent Dev Corp. US patent US 3567118 A, published 2 March 1971.

Soane, D., Offord, D., Ware, W., 2010. Chapter 9. Nanotechnology application in textiles. In: Shulte, J. (Ed.), Nanotechnology: Global Strategies, Industry Trends and Applications. Wiley, India, pp. 149–162.

Steinmann, W., Walter, S., Beckers, M., Seide, G., Gries, T., 2013. Thermal analysis of phase transitions and crystallization. Available from, In: Elkordy, A.A. (Ed.), Polymeric Fibers, Applications of Calorimetry in a Wide Context—Differential Scanning Calorimetry, Isothermal Titration Calorimetry and Microcalorimetry. InTech, Rijeka, ISBN: 978-953-51-0947-1, pp. 277–306. http://www.intechopen.com (accessed 7.7.15).

Thornton, J.I., 1997. The general assumptions and rationale of forensic identification. In: Faigman, D.L., Kaye, D.H., Saks, M.J., Joseph Sanders, J. (Eds.), Modern Scientific Evidence: The Law and Science of Expert Testimony, vol. 2. West Publishing Company, St. Paul.

Tree-Udom, T., Wanichwecharungruang, S.P., Seemark, J., Arayachukeat, S., 2011. Fragrant chitosan nanospheres: controlled release systems with physical and chemical barriers. Carbohydr. Polym. 86, 1602–1609.

Treigienė, R., Musnickas, J., 2003. Solvent pre-treated wool fabric permanent set and physical properties. Fibres Text. 11, 37–40.

Tsuzuki, T., Wang, X., 2010. Nanoparticle Coatings for UV Protective Textiles. Res. J. Text. Appar. 14, 9–20.

Turi, E. (Ed.), 1981. Thermal Characterization of Polymer Materials. second ed., Academic Press, New York.

Wielage, B., Lampke, T.H., Marx, G., Nestler, K., Starke, D., 1999. Thermogravimetric and differential scanning calorimetric analysis of natural fibres and polypropylene. Thermochim. Acta 337, 169–177.

Wunderlich, B., 2005. Thermal Analysis of Polymer Material. Springer-Verlag, Heidelberg.

Yip, P.W., 2014. Analysis of two methods for characterization of flame resistant military fabrics and commercial textile fibers: simultaneous DSC-TGA and pyrolysis GCMS. Technical report NATICK/TR-14/008. US Army Natick Soldier Research, Development and Engineering Center, Natick, MA.

Yuan, Q., Yao, J., Huang, L., Chen, X., Shao, Z., 2010. Correlation between structural and dynamic mechanical transitions of regenerated silk fibroin. Polymer 51, 6278–6283.

Zhou, G., Shao, Z., Knight, D.P., Yan, J., Chen, X., 2009. Silk fibers extruded artificially from aqueous solutions of regenerated *Bombyx mori* silk fibroin are tougher that their natural counterparts. Adv. Mater. 21, 366–370.

Index

Note: Page numbers followed by *f* indicate figures, *t* indicate tables and *s* indicate scheme.

Edwards Brothers Malloy
Ann Arbor MI. USA
June 21, 2016